AF871694

LEED® AP BD+C

V5 EXAM PREP

PRACTICE TESTS & SUMMARY SHEETS

SECOND EDITION

(BUILDING DESIGN & CONSTRUCTION)

A. Togay Koralturk

Projectific

Copyright © 2026 by Projectific, Inc. All rights reserved.

Except as permitted under the Copyright Act of 1976, no part of this publication, including the content and information that can be accessed in the digital resources of this book via QR codes, hyperlinks, or otherwise, may be reproduced or distributed in any form or by any means, or stored in a database or retrieval system, without the prior written permission of publisher.

Published by:
Projectific, Inc.
8 The Green
Suite #4445
Dover, DE 19901
www.projectific.com

ISBN: 978-1-972631-08-9 (paperback)

Printed in the United States of America.

ATTENTION CORPORATIONS, UNIVERSITIES, COLLEGES, AND PROFESSIONAL ORGANIZATIONS. Special quantity discounts are available on bulk purchases of this book. Desk copies are also available for universities and academic institutions. For more information, please contact: customercare@projectific.com

Visit us on the Web: projectific.com

LEED® and USGBC® are registered trademarks of the U.S. Green Building Council. The author and publisher are not affiliated with the U.S. Green Building Council (USGBC) or the Green Business Certification Inc. (GBCI) and does not administer LEED Professional Exams or the LEED green building certification program. The author and publisher do not claim any endorsement or recommendation of its products or services by USGBC or GBCI.

While the publisher and author have used their best efforts in preparing this book, they make no representations or warranties with respect to the accuracy or completeness of the contents of this book, including any content and information that can be accessed in the digital resources of this book via QR codes, hyperlinks, or otherwise, and specifically disclaim any implied warranties of merchantability or fitness for a particular purpose. The advice and strategies contained herein may not be suitable for your situation. Neither the author, publisher nor its licensors shall be liable to you or anyone else for any inaccuracy, error or omission, regardless of cause, in the work or for any damages resulting therefrom. The author and publisher have no responsibility for the content of any information accessed through the work. Under no circumstances shall the author, publisher and/or its licensors be liable for any indirect, incidental, special, punitive, consequential or similar damages that result from the use of or inability to use the work, even if any of them has been advised of the possibility of such damages. No warranty may be created or extended by sales representatives or written sales materials.

All content in this book, including the content and information that can be accessed in the digital resources of this book via QR codes, hyperlinks, or otherwise, is provided as is and is subject to change without notice.

CONTENTS

PART 1 PRACTICE TESTS

PART 2 SUMMARY SHEETS

SECTION 5: SUMMARY SHEETS

ALSO BY A. TOGAY KORALTURK

SECTION 1:

ABOUT THE LEED AP BD+C EXAM

EXAM CONTENT

The LEED AP BD+C exam contains 100 randomly delivered multiple-choice questions and must be completed in 2 hours. The exam questions reflect Knowledge Domains aligned with the LEED v5 rating system credit categories. Unlike the LEED Green Associate exam, which tests general green building knowledge, the LEED AP BD+C exam tests your ability to apply the prerequisites and credits to real projects. GBCI writes the questions at three cognitive levels—recall, application, and analysis.

Below is the breakdown of the exam content areas and the number of questions per category:

- LEED Process (7 questions): Minimum Program Requirements, Platinum requirements, credit interpretation rulings, LEED equivalencies, system synergies and tradeoffs, the project/LEED/property boundaries, addenda and development cycles, rating system selection, and LEED occupancy types (FTE, transients, visitors, residents).
- Integrative Process, Planning, and Assessments (8 questions): The integrative design process (timing, engagement, charrette, discovery, goal setting), core and shell considerations (tenant guidelines, green leases), climate resilience, human impact and social context, and carbon assessment.

- Location and Transportation (8 questions): Development constraints and opportunities, community connectivity, access to quality transit, transportation demand management, electric vehicles, and the economic and social vitality of communities.
- Sustainable Sites (9 questions): Site assessment, construction activity pollution prevention, habitat conservation and restoration, exterior lighting, heat island reduction, rainwater management, extreme climate events, and accessible quality open space.
- Water Efficiency (9 questions): Irrigation demand, fixtures and fittings, appliance and process water, types and quality of water, water reuse, whole-project water use, water use measurement, and leak detection.
- Energy and Atmosphere (18 questions): Energy performance and efficiency, electrification, reduction of peak thermal loads, renewable energy supply and EACs, refrigerants and GWP benchmarks, commissioning, advanced energy metering, energy storage and power resilience, and grid interactivity.
- Materials and Resources (13 questions): Whole-building life-cycle assessment, procurement of low-embodied carbon materials, construction-phase carbon tracking, building and material reuse, material stewardship, low-emitting materials, human and ecological health impacts, and construction, demolition, and operational waste management.
- Indoor Environmental Quality (13 questions): The thermal, sound, lighting, and biophilic environments, accessibility and inclusion, construction management, ventilation levels, indoor air quality management and testing, smoking and vehicle idling, management modes for risk events, and resilient planning and design.

LEED AP BD+C EXAM POLICIES AND PROCEDURES

In this section, we will provide details on the LEED AP BD+C exam policies, procedures, and other important aspects you should know before taking your exam. However, as GBCI's policies and procedures are subject to change, it is also recommended that you visit USGBC's website and review the latest LEED AP with Specialty Candidate Handbook.

HOW TO REGISTER FOR THE EXAM?

To register for the LEED AP BD+C exam, you need to complete the following steps:

1) Create a USGBC site user account at www.usgbc.org/credentials, or log in if you already have one. **When filling out the required information, make sure that the name you enter exactly matches your name on the ID you will present to take the exam**. Any discrepancy may disqualify you from taking the exam.

2) Under the *Credentials* section, select the LEED AP BD+C credential and follow the instructions to complete the application. If your organization is a USGBC member, link your member status to your account before registering to receive member pricing.

3) Once your application is submitted and payment is processed, you will receive an Eligibility ID (EID) and will be redirected to prometric.com/gbci to schedule your exam date and location. Alternatively, you may choose to take the exam online through remote proctoring. Your EID is valid for 12 months from the date of registration.

4) When the exam is scheduled, you will receive a confirmation number onscreen and an email through Prometric. Keep your confirmation number, as you will need it to confirm, cancel, or reschedule your appointment.

WHERE TO TAKE THE EXAM?

The LEED AP BD+C exam can be taken at Prometric testing centers around the world, or you can take the exam online, in the comfort of your home or office, while being monitored by an offsite proctor. In the following sections, we will discuss the policies and procedures for both the testing center and the online proctored test.

TESTING CENTER POLICIES AND PROCEDURES

Test-takers who choose to take their exam at a Prometric testing center should arrive 30 minutes before their scheduled appointment to sign in and present their identification, Eligibility ID (EID), and the confirmation number, which are emailed upon scheduling the appointment. Test-takers who arrive later than their scheduled exam time will lose their seat.

When signing in, test-takers must present a valid, non-expired, government-issued photo ID that also contains a signature. Once more, your name on the identification must exactly match the name on your exam registration. In case of a discrepancy, you may not be permitted to take your exam.

Nothing can be brought to the testing area. Small lockers may be available at the testing center to store your wallet, cell phone, keys, and other small items. However, laptops, briefcases, or large purses should not be taken to the testing center.

Throughout the exam session, you should remain seated. If you experience an issue or need to take a break, raise your hand to notify test center staff. Note that the exam will not pause if you take a break, and you must get the test proctor's permission to leave the room.

Expect to see other individuals entering and exiting the testing area throughout your exam, as Prometric testing centers offer a variety of exams. If the testing center is in a busy area, be prepared for potential noise distractions.

POLICIES AND PROCEDURES FOR ONLINE PROCTORED TEST

To take the LEED AP BD+C exam online, you will need a computer equipped with a webcam, a stable internet connection, and a private, quiet space where no one will enter or pass through your testing area.

Before scheduling your exam, be sure to review the online proctoring requirements on the Prometric website at prometric.com/gbci. On that website, you will have the opportunity to run a system test to ensure your computer and internet connection meets all the necessary criteria for a smooth testing experience. Taking this step before scheduling your exam is crucial to avoid any technical issues that could interrupt or invalidate your exam session. If your system cannot pass the test, you may instead choose to take the exam at a testing center.

On exam day, you will need to check in for the online proctored test before your scheduled appointment. During the check-in process, you will need to run a system check and provide photos of yourself, your identification, and your testing space. Your identification must be a valid, non-expired, government-issued photo ID that also contains a signature.

Once you submit your photos, they will be reviewed by the offsite proctor. If there is any issue, the proctor will contact you. To speed up the check-in process, make sure that your desk and the area around you are clear of all personal and preparatory items, including books, notes, and other materials. Otherwise, the proctor may ask you to show all sides of your testing environment via your webcam.

Throughout your online exam session, you will be under surveillance via your webcam and microphone by the offsite proctor. If someone enters your testing space or if there are loud distractions, you risk having your exam terminated. For this reason, choose a testing space free of interruptions and background noise.

Once everything meets the requirements, the proctor will start your exam. You should remain seated in front of your webcam throughout the exam session. If you encounter any technical issue during your exam, the proctor will attempt to resolve it in real time.

SPECIAL TESTING ACCOMODATIONS

Accommodations can be requested if you have a documented disability that would prevent you from taking the exam under normal testing conditions. Prometric complies with the provisions of the Americans with Disabilities Act (ADA). There is no additional charge for special accommodations.

DURING THE EXAM

EXAM TUTORIAL AND EXIT SURVEY

Before the exam, there will be a 10-minute tutorial to demonstrate the exam software. At the end of the exam, there will be a 10-minute optional exit survey. The time allocated for the tutorial and the survey is not included in your 2-hour examination time.

THE EXAM

The LEED AP BD+C exam contains 100 randomly delivered multiple choice questions that must be completed in 2 hours.

As the exam starts, you will be presented with one question at a time. For every question, you have the option to answer the question, answer and mark it for later review, or skip it. You have the flexibility to move back and forth among the questions at any point and change your answer to any question you have previously addressed. The exam software will also list all the questions you have either skipped or marked for review so that you can easily revisit them.

Make sure you read all the questions and the choices very carefully. If a question seems to have more than one answer, make sure you thoroughly understand the question and pay special attention to the wording. It doesn't matter how well you know the exam content; if you don't read carefully, the exam can easily trick you into selecting the wrong answer.

Answer every single question since leaving a question blank will not award any points. If you are unsure about the answer to a question and are likely to spend extra time deciding on the correct answer, mark the question and come back to it later. Often, you will find it easier to answer such questions upon returning to them with a fresh perspective. If you forget something and struggle to answer a question, the questions or answer choices you see later in the exam can help you remember the right answer to a previous question.

At the end of the exam, make sure you use any extra time to review your answers, starting with the questions you felt most unsure about.

AFTER THE EXAM

EXAM RESULTS

After the exam, your exam score will be displayed on screen, and a score report will follow. Your results are then processed to your Credentials Dashboard, where you can download your certificate as a PDF.

Note that candidates who take the LEED v5 beta exam will not receive immediate results. Instead, they receive a Notice of Completion, and scores are released after the beta period concludes and the psychometric analysis is complete.

PASSING THE EXAM

A score of 170 or higher earns the credential. As soon as you receive written confirmation that you have passed, you may use "LEED AP BD+C" after your name. Note that your LEED Green Associate credential expires, as the LEED AP BD+C credential supersedes it.

FAILING THE EXAM

The full exam fee is charged for each attempt. After three unsuccessful attempts within 12 months, you must wait 90 calendar days before submitting a new registration to GBCI.

HOW TO MAINTAIN YOUR CREDENTIAL

LEED APs with specialty must renew their credential every 2 years by earning and reporting 30 continuing education (CE) hours within the reporting period, including 6 LEED-specific hours. The reporting period starts on your exam date.

SECTION 2:

PRACTICE TEST 1

It is recommended to study the Summary Sheets and Glossary before taking the practice tests.

1) Under the Planning for Zero Waste Operations prerequisite, what must every LEED project provide?

a) A recycling program
b) An on-site composting facility
c) A waste-to-energy contract
d) A zero-landfill guarantee

2) Under the Water Metering and Reporting prerequisite, a LEED BD+C: Core and Shell project is a speculative warehouse shell delivered with no installed plumbing fixtures, fittings, or water-using systems in the project scope. With respect to the water metering prerequisite, how is this project treated?

a) It must still install a whole-building meter sized for anticipated future tenant demand
b) It is exempt from the prerequisite
c) It must install a minimum of one meter per floor for future tenant spaces
d) It may defer all metering until the first tenant fit-out is permitted

3) In its submittal for the Climate Resilience Assessment prerequisite, a team states that it applied an Intergovernmental Panel on Climate Change emissions scenario when evaluating each priority hazard. What additional specification does the prerequisite require regarding that scenario?

a) The general circulation model and grid resolution used to downscale the projection

b) The shared socioeconomic pathways applied

c) The return period, in years, associated with each modeled hazard event

d) Written concurrence from the local authority having jurisdiction on the scenario chosen

4) A team selecting the adaptable environment option of Occupant Experience asks what the seating strategy requires. What does the listed strategy provide?

a) Reserved bariatric seating in 5% of positions

b) Standing desks for all occupants

c) Identical ergonomic chairs at every position

d) Choice in furniture configuration and a variety of seating accommodating a wide range of body types

5) A New Construction project wants to pursue the optimize water use for cooling path of the optimize process water use option of the Enhanced Water Efficiency credit. Which condition must be satisfied for the project to be eligible for this path?

a) The baseline system designated for the building using ASHRAE Standard 90.1, Appendix G must include a cooling tower

b) The project's process water use must represent at least 10% of total building regulated water use, excluding cooling

c) The project must conduct a one-time potable water analysis measuring at least five control parameters

d) The project must use a minimum of 20% alternative water for the cooling system

6) Under Enhanced Commissioning, Option 2, Path 2 (Enhanced Software), a project must provide a remotely accessible fault detection and diagnostics (FDD) system. Above what total project installed capacity of cooling, heating, or refrigeration systems is this FDD requirement triggered?

a) 3,600 kBtu/hr (300 tons or 1,055 kW)

b) 7,200 kBtu/hr (600 tons or 2,110 kW)

c) 10,000 kBtu/hr (833 tons or 2,930 kW)

d) 14,400 kBtu/hr (1,200 tons or 4,220 kW)

7) Under the Fundamental Air Quality prerequisite, which reference standards are used to achieve acceptable indoor air quality?

a) ASHRAE 62.1-2022 and 62.2-2022

b) ASHRAE 90.1-2022 and 55-2020

c) ASHRAE 52.2-2017 and 241-2023

d) ASHRAE 189.1 and 100

8) Under Construction and Demolition Waste Diversion, how does the credit treat the combustion of recovered wood materials from the construction waste recycling process (wood-derived fuel or biomass) relative to other forms of waste-to-energy?

a) It is treated identically to all other incineration and counts as zero percent diverted

b) It is an acceptable means of diversion and is not considered waste-to-energy for LEED project diversion reporting, unlike all other incineration

c) It may be counted as diversion only if the facility holds third-party verification of its recycling rate

d) It is capped at a maximum 35 percent diversion rate like unverified mixed materials

9) A hospital pursues the power resilience option of the Grid Interactive credit. The design pairs on-site renewable generation with energy storage so the building can island and power its critical equipment independently from the grid. For at least how long must the project be able to power those critical loads?

a) At least three days

b) At least 24 hours

c) At least 48 hours

d) At least one week

10) A school project pursues the equitable access to resources option of the Equitable Development credit and wants both of its paths. Which combination correctly states the minimum number of space types required for the public use spaces path and the community partnership path?

a) At least two shared-use space types for public use spaces, and at least three dedicated-use space types for community partnership

b) At least three shared-use space types for public use spaces, and at least three dedicated-use space types for community partnership

c) At least two shared-use space types for public use spaces, and at least two dedicated-use space types for community partnership

d) At least three shared-use space types for public use spaces, and at least two dedicated-use space types for community partnership

11) A team designing for the limit-refrigerant-leakage option of the Enhanced Refrigerant Management credit must minimize field-installed piping. The credit requires self-contained equipment with no field-installed piping in which two situations?

a) For equipment with refrigerants greater than 300 GWP, and for at least 50% of the total GWP of refrigerants used in the project

b) For all equipment regardless of GWP, and for 100% of the total refrigerant GWP

c) For equipment with refrigerants greater than 1,400 GWP, and for at least 80% of the total refrigerant charge by weight

d) For equipment with refrigerants greater than 700 GWP, and for at least 80% of the total GWP of refrigerants used in the project

12) Under the Renewable Energy credit, where may Tier 1 renewable generation equipment be located?

a) Anywhere within the same country as the project

b) On the project site, on the campus where the project sits, or on the site of a social impact project

c) Only physically on the project's own roof

d) At any Green-e Energy certified off-site generation facility

13) Under the Biodiverse Habitat credit, a team pursuing vegetation restoration under the preserve-and-restore option is selecting plant species for a previously disturbed site in the U.S. Aside from avoiding invasive species, what is the minimum number of native or adapted species required, and to what must they be native or adapted?

a) At least 10 species native or adapted to the project's EPA Level III ecoregion

b) At least 8 species native or adapted to the project's EPA Level III ecoregion

c) At least 10 species native or adapted to the project's EPA Level IV ecoregion

d) At least 12 species native or adapted to the project's USDA plant hardiness zone

14) Under Building and Materials Reuse, Option 2: Materials Reuse, which of the following is one of the listed targeted materials that receives a 2x multiplier rather than being categorized as an "other" material?

a) Dimensional lumber

b) Ceilings

c) Doors

d) Plumbing fixtures

15) A team is assigning products to the correct low-emitting materials categories. According to the credit's category definitions, which assignment is correct?

a) Foamed-in-place insulation sprayed on-site is documented under the paints and coatings category

b) An interior door is documented under the walls category

c) A countertop is documented under the walls category

d) A wet-applied product applied on the floor is documented under the flooring category

16) A LEED BD+C: Core and Shell project addresses drought as one of its two highest-priority hazards under Enhanced Resilient Site Design and includes a created water feature. Which requirement applies specifically to makeup water for that water feature?

a) Makeup water must not exceed 10,000 gallons (37,854 liters) of potable water per year.

b) Makeup water must come 100% from harvested rainwater with no potable water permitted.

c) Makeup water must not exceed 5,000 gallons (18,927 liters) of potable water per year, or at least 75% of annual makeup water must come from alternative water sources.

d) Makeup water must not exceed 5,000 gallons (18,927 liters) of potable water per year, with no alternative-source option allowed.

17) During the design phase of Fundamental Commissioning, the commissioning provider must attend design meetings. What is the minimum attendance?

a) At least one meeting focused on mechanical, electrical, and plumbing and one focused on the envelope, which may be separate or combined
b) Every design coordination meeting held by the architect
c) One kickoff meeting at schematic design only and one meeting focused on the envelope
d) One meeting during the design phase and one meeting during the construcion phase

18) Under the location-efficient affordable housing path of the Equitable Development credit, a residential project relies on the jobs-to-housing ratio condition rather than the transit or walkable-location conditions. What jobs-to-housing ratio must the community exceed, and within what walking distance?

a) A ratio exceeding 1:2 within 0.25 miles (400 meters) walking distance
b) A ratio exceeding 1:1 within 0.5 miles (800 meters) walking distance
c) A ratio exceeding 1:2 within 0.5 miles (800 meters) walking distance
d) A ratio exceeding 2:1 within 0.5 miles (800 meters) walking distance

19) A project in a linguistically diverse neighborhood pursues the signage strategy in the social health category of Accessibility and Inclusion. What does the strategy require?

a) Signage in English plus the single most common other language
b) Signage in all languages spoken by more than 10% of the local population
c) Pictogram-only signage to avoid language choices
d) Signage in all languages spoken by more than 5% of the local population

20) A project addresses high winds under the Enhanced Resilient Site Design credit in an area that is not hurricane-prone. Which standard does the credit reference for wind design?

a) FEMA 543
b) The FORTIFIED Commercial Wind standards
c) ASCE 24
d) ASCE/SEI 7

21) USGBC generates a business-as-usual (BAU) carbon projection for every LEED BD+C project under the Operational Carbon Projection and Decarbonization Plan prerequisite. Over what period does the projection run, and what assumption does USGBC make about the electrical grid's emissions factor?

a) A 10-year period during which the grid emissions factor is held constant at the current national coefficient
b) A 25-year period during which the grid emissions factor declines in a straight line by 95%
c) A 30-year period during which on-site combustion emissions decline to zero while grid emissions remain constant
d) A 25-year period during which the grid emissions factor declines in a straight line by 90%

22) A LEED BD+C: New Construction project pursues Enhanced Energy Efficiency, Option 2, Energy Simulation, Path 1 (excluding on-site renewable contribution). What percentage improvement in the performance index below the performance index target is required to earn the maximum 10 points on Path 1?

a) 100%

b) 27%

c) 30%

d) 24%

23) In the LEED point allocation process, points are allocated to each credit according to a credit's contribution to the:

a) LEED guiding principles

b) LEED impact areas

c) LEED credit categories

d) LEED priority list

24) Under the Accessibility and Inclusion credit, a project is located in a jurisdiction with no applicable accessibility code identified during the Human Impact Assessment. Beyond providing ramps for accessible routes and a front-approach wheelchair-accessible counter section, how many additional foundational accessibility strategies must be demonstrated, and what is the third one?

a) Two additional strategies: clear width for human passage and accessible parking

b) One additional strategy: audible and visual emergency alerts

c) Two additional strategies: clear width for human passage and nonslip flooring

d) One additional strategy: clear width for human passage

25) Mandatory requirements are unevenly distributed across the LEED v5 BD+C categories. Which credit category holds the greatest number of prerequisites?

a) Water Efficiency, with two

b) Energy and Atmosphere, with five

c) Materials and Resources, with two

d) Indoor Environmental Quality, with three

26) A team conducting the whole-building life-cycle assessment for Reduce Embodied Carbon, Option 1 is setting up its baseline and proposed models. Which of the following is not a requirement for a compliant WBLCA under this option?

a) The assessment scope must be cradle-to-grave (modules A–C), excluding operating energy and operating water-related energy

b) Material quantities must not change by more than 5% from design through construction

c) The service life of the baseline and proposed buildings must be the same and at least 60 years

d) The baseline and proposed buildings must be of comparable size, function, orientation, and operating energy performance

27) A demolition crew salvages and reuses heavy timber on-site in the new building, and the project team wants to count its full value. Under the construction and demolition waste diversion credit, how should these on-site reused materials be handled in the diversion calculation?

a) Counted at 200% of their weight because reuse is prioritized over recycling

b) Excluded from this credit's diversion calculation; they instead count toward Building and Materials Reuse

c) Counted at 100% of their weight as source-separated material

d) Counted at the facility average diversion rate for the demolition phase

28) Under the Occupant Experience credit, within the quality electric lighting path, a designer chooses to demonstrate acceptable discomfort glare for each space using the unified glare rating rather than per-luminaire luminance. What is the maximum unified glare rating permitted?

a) 19 or lower

b) 16 or lower

c) 22 or lower

d) 25 or lower

29) A team pursuing the parking-under-cover option for heat island reduction is determining how many of its parking spaces must be placed under cover to earn the point. What percentage of parking spaces must be covered?

a) 75% of parking spaces

b) 80% of parking spaces

c) 100% of parking spaces

d) 90% of parking spaces

30) Under Option 1 of the Enhanced Commissioning credit, what do the two paths cover?

a) Path 1 covers monitoring-based commissioning, and Path 2 covers envelope commissioning

b) Path 1 provides enhanced commissioning for MEP systems, and Path 2 additionally provides building enclosure commissioning

c) Path 1 covers the envelope, and Path 2 covers renewable energy systems only

d) Path 1 covers fundamental commissioning, and Path 2 covers monitoring software

31) A Core and Shell team is comparing Option 1 and Option 2 of the Green Leases credit. Which statement correctly describes what each option rewards?

a) Option 1 rewards executing a lease with a signed tenant, while Option 2 rewards drafting the standard lease language

b) Option 1 rewards developing a standard green lease and committing to incorporate it into future tenant leases, while Option 2 rewards actually executing that lease with a tenant

c) Option 1 rewards earning third-party Green Lease Leaders recognition, while Option 2 rewards drafting the standard lease

d) Both options require a fully executed lease; they differ only in the number of tenants covered

32) A team pursuing the whole-project water use option of Enhanced Water Efficiency asks what minimum reduction its proposed model must show. What is the entry threshold?

a) A 25% reduction from the project's baseline

b) A 40% reduction from the project's baseline

c) A 30% reduction from the project's baseline

d) A 20% reduction from the project's baseline

33) Under Option 2 of the Green Leases credit, a tenant has not yet finalized the green lease but the team still wants the additional point. Which of the following may the tenant submit as the permitted alternative to the executed green lease?

a) A verbal commitment recorded in the design charette meeting minutes

b) A draft lease initialed by the landlord only

c) A memorandum of understanding signed by the broker on the tenant's behalf

d) A signed and dated letter of attestation confirming the tenant's commitment to the sustainability criteria in the green lease

34) A project team is pursuing LEED v5 BD+C: New Construction Platinum certification. The project has earned 85 points, is fully electric with no on-site combustion, procures 100% renewable energy through a combination of on-site solar and Tier 3 RECs, and has achieved a 20% reduction in embodied carbon. However, the project earned only 6 points in the Enhanced Energy Efficiency credit. Will this project achieve Platinum?

a) Yes, because 85 points exceeds the 80-point Platinum threshold

b) Yes, because the project meets three of the four decarbonization requirements

c) No, because New Construction Platinum requires 8 points in the Enhanced Energy Efficiency credit

d) No, because Tier 3 RECs do not count toward the 100% renewable energy requirement

35) Under the Operational Carbon Projection and Decarbonization Plan prerequisite, who must review the carbon projection and the decarbonization plan, and what must they do?

a) The commissioning authority, who must attest that they have reviewed them

b) The energy modeler, who must stamp the results

c) The building owner or the owner's representative, who must attest that they have reviewed them

d) The local authority having jurisdiction, which must approve the projection before submittal

36) During construction, the commissioning provider on a New Construction project performs a quality assurance/quality control review of the contractor's completed commissioning documentation. To satisfy the minimum required by the Fundamental Commissioning prerequisite, at least what fraction of those completed documents must the provider review?

a) At least 5%

b) At least 15%

c) At least 10%

d) At least 25%

37) Under the light pollution reduction credit, certain facade and landscape lighting is exempt from the uplight and light trespass requirements in higher lighting zones, but only under a specific operating condition. Which condition must this lighting meet to qualify for the exemption?

a) It is in lighting zones 2 and 3 and automatically turned off from 11 p.m. until 5 a.m.
b) It is in lighting zones 3 and 4 and automatically turned off from midnight until 6 a.m.
c) It is in lighting zones 3 and 4 and automatically turned off from 10 p.m. until dawn
d) It is in lighting zones 1 and 2 and automatically turned off from midnight until 6 a.m.

38) Under the Transportation Demand Management credit, the Shower and Changing Facilities path requires providing access to on-site showers with changing facilities for what proportion of all regular building occupants?

a) 0.5% of all regular building occupants
b) 1% of all regular building occupants
c) 2.5% of all regular building occupants
d) 5% of all regular building occupants

39) A team is preparing the submittal for an innovation strategy under the Project Priorities credit. The credit requires the team to identify several specific criteria for the proposed innovation strategy. Which of the following is not one of the required criteria?

a) A third-party verification of the achieved environmental benefit
b) The intent of the proposed innovation strategy
c) Proposed requirements for compliance
d) Proposed submittals to demonstrate compliance

40) A New Construction project sits on a site that also contains a detached parking garage and a small outbuilding, both inside the LEED project boundary. The team proposes excluding these two structures from the embodied carbon quantification because they are not part of the occupied building. Under the Quantify and Assess Embodied Carbon prerequisite, how should these ancillary structures be treated?

a) They may be excluded because they are detached from the main occupied building
b) They must be included in the calculations because they are within the LEED project boundary
c) Only the parking garage must be included; outbuildings are always optional
d) They may be excluded if their combined embodied carbon is less than 10% of the project total

41) Under the Enhanced Energy Efficiency credit, which of the following is an example of process energy, also called unregulated energy?

a) Interior and exterior lighting
b) HVAC and service hot water heating
c) Computers, kitchen refrigerators and stoves, washers and dryers, and elevators and escalators
d) Parking garage ventilation and toilet exhaust

42) A new commercial office building includes a surface parking lot with 125 spaces and pursues Electric Vehicles, Option 1, targeting 1 point. Using the credit's rounding rule, how many total EVSE-equipped spaces must the project provide, and what additional condition must be satisfied?

a) 6 spaces, with no accessibility requirement
b) 7 spaces, with no accessibility requirement
c) 7 spaces, including at least one in an accessible parking space
d) 13 spaces, including at least one in an accessible parking space

43) On a project pursuing the construction management prerequisite, the mechanical contractor determines that the permanently installed air handlers must be operated during the final weeks of construction to condition the space. Which action is required so that running this equipment still complies with the prerequisite's HVAC protection criterion?

a) Operate the equipment only on a 100% outdoor-air economizer cycle during construction
b) Maintain proper filtration on the equipment while it is in use during construction
c) Limit equipment runtime so that it operates no more than eight hours per day
d) Bypass the return-air path and pressurize the building to keep contaminants out

44) Under the Grid Interactive credit, Option 1, Energy Storage, how must the storage operate?

a) It provides on-site electric or thermal storage with automatic load management controls that store energy during off-peak periods or periods of low grid carbon intensity and deploy it during on-peak periods or periods of high grid carbon intensity
b) It stores energy continuously and discharges only during a utility-declared emergency
c) It provides thermal storage only, discharged on a fixed daily schedule
d) It stores surplus on-site renewable generation for export back to the grid

45) A project team installs a permanent landscape irrigation system and pursues the Minimum Water Efficiency prerequisite via Minimum Outdoor Water Use Efficiency, Option 2. Efficient Irrigation. What is the minimum reduction the installed system must achieve, and against what baseline?

a) At least 20% reduction from the calculated baseline for the site's annual theoretical irrigation requirement (TIR)
b) At least 20% reduction from the EPA WaterSense baseline
c) At least 30% reduction from the calculated baseline for the site's annual theoretical irrigation requirement (TIR)
d) At least 40% reduction from the project's metered prior-year irrigation use

46) A New Construction team installs a geothermal heat pump that uses a vapor-compression cycle and expects to count its output as Tier 1 on-site renewable energy. A LEED AP corrects the team. How is this system treated?

a) It qualifies as Tier 1 on-site renewable energy because the ground loop is located on the project site.

b) It qualifies as Tier 3 off-site renewable energy because the subterranean heat originates outside the building.

c) It qualifies as Tier 1 only if the recovered heat is drawn from municipal wastewater rather than the ground.

d) It is not a renewable energy resource.

47) A LEED AP is comparing the two compliance paths of Rainwater Management. Which statement correctly distinguishes Option 2: Natural Land Cover Conditions from Option 1: Percentile of Rainfall Events?

a) Option 2 requires retaining the runoff from the 90th percentile rainfall event.

b) Option 2 requires a minimum 30-year record of daily rainfall, whereas Option 1 uses only a 10-year record.

c) Option 2 requires retaining on-site the increase in runoff volume relative to pre-disturbance natural land cover.

d) Option 2 awards 1 additional point for collecting and reusing rainwater, just as Option 1 does.

48) A project pursuing the Light Pollution Reduction credit illuminates a national flag on its site. Under what condition is that lighting exempt?

a) In any lighting zone, without restriction

b) In lighting zones 2, 3, or 4

c) Only in lighting zone 0, where flag lighting is the sole permitted source

d) Only if the flag lighting is extinguished from midnight until 6 a.m.

49) After volunteer committees develop proposed changes to LEED, what is the correct sequence of approvals before the changes are published?

a) USGBC membership votes first, then the Board of Directors reviews, then the LSC approves

b) The LSC and Board of Directors review and affirm, then the consensus committee votes, then USGBC membership ratifies

c) GBCI reviews and approves independently without committee involvement

d) The consensus committee publishes changes directly without further review

50) A New Construction project used the indoor air quality procedure (IAQP) to comply with Fundamental Air Quality. The team now wants to earn Enhanced Air Quality. Which option(s) are available to them, and what is the Option 2 design limit for PM2.5?

a) Only Option 1 (Increased Ventilation) is available; PM2.5 limit of 12 μg/m3.

b) Either Option 1 or Option 2 is available; PM2.5 limit of 35 μg/m3.

c) Only Option 2 (Enhanced Indoor Air Quality Design) is available; PM2.5 design limit of 10 μg/m3.

d) Only Option 2 (Enhanced Indoor Air Quality Design) is available; PM2.5 design limit of 20 μg/m3.

51) A project provides all-gender restrooms under the social health category of Accessibility and Inclusion. What does the strategy specify?

a) A single all-gender restroom in the lobby only

b) Conversion of every restroom to all-gender

c) At least one fully accessible, all-gender, single-use restroom, or one multiuse all-gender restroom, on each floor of the building

d) All-gender restrooms only on floors with more than 100 occupants

52) A project team operated the permanently installed HVAC system during construction with temporary filters in place. As construction wraps up, the team plans the sequence of close-out activities to satisfy the construction management prerequisite. Regarding the air filtration media and the testing and balancing (TAB) work, which sequence is required?

a) Complete testing and balancing first, then replace all air filtration media before occupancy

b) Replace air filtration media only if the temporary filters fall below the design rating, with no TAB resequencing

c) Perform testing and balancing before occupancy but retain the construction-phase filters to preserve calibration

d) Replace all air filtration media after construction is complete, then perform testing and balancing with the new filters

53) Under the Electrification credit, under the space-heating path of the New Construction low-temperature combustion option, the requirement to achieve a weighted average space heating equipment efficiency of at least 1.8 coefficient of performance applies only to projects in certain climate zones. For which climate zones is this efficiency requirement mandatory?

a) Climate zones 0 through 2

b) Climate zones 4 and above

c) All climate zones without exception

d) Climate zones 3 and above

54) A commercial building rejects heat from several appliances and is documenting the process standards of the Minimum Water Efficiency prerequisite. Which requirement applies to heat rejection and cooling for that equipment?

a) Once-through cooling with potable water is permitted if the discharge is captured for irrigation

b) No once-through cooling with potable water is allowed for any equipment or appliances that reject heat

c) Once-through cooling with potable water is allowed provided a makeup water meter is installed

d) Once-through cooling with potable water is allowed only for medical equipment under emergency backup

55) An architect pursuing the bird-friendly glass option of Biodiverse Habitat argues that the glass infill of a rooftop terrace guardrail, located 60 feet (18 meters) above grade, is exempt because the option only regulates glass up to 50 feet (15 meters). How should a LEED AP respond?

a) The guardrail glass must still comply, because glass in guardrails and windshields is regulated at any distance from grade or roof

b) The architect is correct; all glass above 50 feet (15 meters) is categorically exempt from the threat factor limit

c) The guardrail glass is exempt only if it is spandrel glass with a threat factor above 30

d) The guardrail glass must comply only if the terrace is a green roof

56) In establishing project occupancy counts, how does LEED v5 classify higher-education classroom students, and how does this differ from K–12 students?

a) Higher-education classroom students are regular occupants; K–12 students are visitors

b) Higher-education classroom students are visitors; K–12 students are regular building occupants

c) Both are regular building occupants

d) Both are visitors

57) During LEED goal setting under the Integrative Design Process credit, a team defines specific and measurable goals across the three LEED v5 impact areas. The credit requires that these goals be formally embedded in a particular owner-facing project document. Into which document must the goals be incorporated?

a) The basis of design

b) The construction documents

c) The owner's project requirements

d) The commissioning plan

58) A New Construction project pursues the Water Metering and Leak Detection credit by installing submeters. For the indoor plumbing fixtures and fittings subsystem, the meters must serve at least what percentage of the indoor fixtures and fittings addressed in the minimum water efficiency prerequisite?

a) 80%
b) 50%
c) 75%
d) 100%

59) Under the Occupant Experience credit, a team pursues the 2-point acoustic criteria path of the sound environment option for a school building. Beyond satisfying all classrooms and other core learning spaces, what minimum percentage of the occupied spaces must comply with the acoustic criteria defined during the mapping exercise?

a) at least 90% of the occupied spaces
b) at least 60% of the occupied spaces
c) at least 75% of the occupied spaces
d) at least 80% of the occupied spaces

60) Under the Building and Materials Reuse credit, Option 1, what must the project maintain?

a) The existing building structure and enclosure
b) The interior finishes and furnishings
c) The mechanical and electrical systems
d) The site hardscape and landscaping

61) A project keeps part of an existing structure in place and also procures salvaged structural steel from another site, incorporating it into the new building's frame. How does the building reuse option of Building and Materials Reuse treat the procured steel?

a) It counts only if the steel comes from a building on the same campus
b) It counts; the option addresses reuse of structure and enclosure materials left in situ as well as those procured off-site and incorporated into the building's structure or enclosure
c) It counts only under Construction and Demolition Waste Diversion
d) It does not count; only materials original to the site qualify for building reuse

62) Under Building Product Selection and Procurement, a product's documentation earns the highest of the three achievement levels within a criteria area, described as elite and representing the forefront of sustainability. Which level is this, and what multiplier applies?

a) Level 1, with a 1x multiplier
b) Level 2, with a 2x multiplier
c) Level 3, with a 5x multiplier
d) Level 3, with a 3x multiplier

63) A project team is establishing outdoor smoking restrictions for a commercial office building pursuing No Smoking or Vehicle Idling. According to the prerequisite, designated outdoor smoking areas must be located at least 25 feet (7.5 meters) from which of the following?

a) The property line and all public sidewalks adjacent to the site.

b) Only the building's primary exterior entrances.

c) All entries and operable windows.

d) All entries, outdoor air intakes, and operable windows.

64) A project team locates near an existing bicycle network to earn the Bicycle Network and Storage path of the Transportation Demand Management credit. What is the maximum allowable walking or bicycling distance from a functional entry and/or bicycle storage to that network?

a) 300 feet (90 meters)

b) 600 feet (180 meters)

c) 0.25 miles (400 meters)

d) 0.5 miles (800 meters)

65) Under Option 1 of the minimum energy efficiency prerequisite, which version of ASHRAE Standard 90.1 applies and what addendum must the project apply?

a) ASHRAE 90.1-2016 with Addendum cr

b) ASHRAE 90.1-2019 with Addendum ch

c) ASHRAE 90.1-2022 with Addendum k

d) ASHRAE 90.1-2019 with Addendum cr

66) A four-story LEED BD+C: Core and Shell office building pursues the submeters option of the Water Metering and Leak Detection credit. Beyond the base submetering requirements, what additional metering must be provided for the future tenant spaces?

a) Meters sufficient to capture total potable water use, with a minimum of one per tenant lease line

b) A single building-level meter capturing total potable use for all future tenant spaces combined

c) Meters sufficient to capture total potable water use, with a minimum of one per floor

d) Meters for at least 80% of the future tenant fixtures, installed at fit-out

67) A LEED BD+C: Core and Shell project wants to earn points under Resilient Spaces. Which option is not available to a Core and Shell project?

a) Option 3. Design for Occupant Thermal Safety During Power Outages, Path 1. Consider Extreme Heat.

b) Option 4. Operable Windows.

c) Option 2. Management Mode for Respiratory Diseases.

d) Option 3. Design for Occupant Thermal Safety During Power Outages, Path 2. Consider Extreme Cold.

68) Under the Reduce Peak Thermal Loads credit, a LEED BD+C: New Construction project pursues the infiltration and balanced ventilation option. During construction, the commissioning team conducts a testing, adjusting, and balancing report of the building air handling systems to verify balanced ventilation on a whole-building basis. The measured ratio of ventilation airflow to exhaust airflow must fall within which range to demonstrate compliance?

a) Between 80% and 120%

b) Between 85% and 115%

c) Between 90% and 110%

d) Between 95% and 105%

69) A team beginning its climate and natural hazard assessment under Climate Resilience Assessment asks which time horizons of hazards must be identified. What does the prerequisite require?

a) Projected and future hazards only, drawn from historical records

b) Observed and projected hazards that could affect the project site and building function

c) Observed hazards only, drawn from historical records

d) Observed, projected, and future natural hazards that could affect the project site and building function

70) A Core and Shell project pursues only Renewable Energy Readiness for 1 point. The design designates a single rooftop solar zone. To comply with the solar readiness criteria, the dedicated solar zone area must equal at least what fraction of the gross roof area?

a) 30% of the gross roof area

b) 20% of the gross roof area

c) 40% of the gross roof area

d) 50% of the gross roof area

71) Under the Heat Island Reduction credit, Option 1: Nonroof and Roof, a team applies the weighted calculation. What must the weighted sum of the site design strategies achieve?

a) It must be greater than or equal to the sum of the total site paved area and the total roof area

b) It must be greater than or equal to half of the total site paved area

c) It must equal the total roof area only, since paving is addressed separately

d) It must be greater than or equal to the total project site area, including landscaped areas

72) A New Construction project has already satisfied Path 1 (paints and coatings, flooring, and ceilings) under the low-emitting materials credit and now wants the full two points. The team plans to add the adhesives and sealants category plus the insulation category. What threshold must each of these two additional categories meet, and how many points does this earn?

a) More than 90% of each additional product category, earning 2 points total

b) More than 80% of each additional product category, earning 3 points total

c) More than 80% of each additional product category, earning 2 points total

d) More than 90% of each additional product category, earning 1 additional point on top of Path 1

73) A project site is in a jurisdiction with no legally adopted flood hazard map and no legal floodplain designation. Under the previously undeveloped sites option of Sensitive Land Protection, where must the development footprint be located with respect to flood risk?

a) At least 100 feet (30 meters) from the edge of any mapped historic flood extent

b) Outside any floodplain subject to a 0.2% or greater annual chance of flooding (500-year floodplain)

c) Entirely outside any floodplain subject to a 1% or greater chance of flooding in any given year

d) Above the design flood elevation plus recommended freeboard

74) Under the Occupant Experience credit, Option 1, Path 2, viewed nature, landmarks, or objects must be located at least how far from the glazing?

a) 25 feet (7.5 meters)

b) 15 feet (4.6 meters)

c) 10 feet (3 meters)

d) 50 feet (15 meters)

75) When preparing the practical recommendations within the tenant guidelines, against what must the recommended sustainable strategies for tenants be aligned?

a) The LEED prerequisites and credits pursued by the base building.

b) The minimum requirements of the local energy code, regardless of the base building's pursued credits.

c) A standardized USGBC tenant checklist applicable to all Core and Shell projects.

d) The certification level the base building ultimately intends to achieve.

76) A team pursuing the integrated biophilic design path of Occupant Experience must demonstrate five principles adapted from Kellert and Calabrese. Which of the following is one of those principles?

a) Biophilic design requires that all materials be biodegradable

b) Biophilic design requires repeated and sustained engagement with nature

c) Biophilic design requires daylight autonomy of at least 55%

d) Biophilic design requires a living green wall in every lobby

77) Which of the following is NOT one of USGBC's six strategic goals?

a) Advocate for green building policies at all levels of government

b) Advance green building globally through knowledge sharing and regional collaboration

c) Strengthen USGBC's internal structure to support market transformation

d) Encourage LEED certification for all new construction projects worldwide

78) Under Compact and Connected Development, Option 3: Walkable Location, a project documents proximity to publicly available uses within 0.5 miles (800 meters). Five restaurants, two supermarkets, one bank, and one pharmacy are all within walking distance. Which restriction would limit how these uses are counted toward the credit?

a) A use may be counted under as many use types as it reasonably fits

b) Counted uses must represent at least four of the five categories

c) No more than two uses in each type of use may be counted

d) All uses within 0.25 miles (400 meters) must be excluded from the count

79) A project team is locating dedicated recycling collection and storage areas to satisfy Planning for Zero Waste Operations. At the time of occupancy, the municipality does not yet offer any organics/food waste collection service to the building. How must the team treat the organics/food waste stream?

a) Organics may be omitted from the design until the municipal service becomes available

b) Space for organics/food waste recycling must still be provided even though service is not yet available

c) Organics must be commingled with mixed paper because no dedicated stream is required

d) Organics may be substituted with an anaerobic digestion contract in lieu of any dedicated space

80) A downtown project has no owned or leased parking; occupants use street parking and public garages. May the project pursue Electric Vehicles?

a) No; projects that do not own or lease parking are not eligible for the credit

b) Yes, by installing EVSE at the nearest public garage

c) Yes, if the municipality operates curbside chargers within 600 feet (180 meters)

d) Yes, through an EV-readiness commitment letter

81) A team seeks Level 1 achievement in the climate health criteria area of the Building Product Selection and Procurement credit for a procured material. What must the product documentation demonstrate to qualify for Level 1 in climate health?

a) A reduction in embodied carbon versus an EPD comparative assessment

b) Quantified and disclosed embodied carbon impacts

c) Disclosed product ingredients to the public or a trusted third party

d) Incorporation of recycled or qualified bio-based content

82) Under the Enhanced Refrigerant Management credit, Option 1, Path 2 is measured against the benchmarks on what basis?

a) Per individual unit

b) Weighted average across all new equipment

c) Highest-GWP unit only

d) Total refrigerant charge by mass

83) In order to be eligible for a LEED BD+C rating system certification, a project should contain a minimum of:

a) 1,000 square feet (93 square meters) gross floor area
b) 2,000 square feet (186 square meters) gross floor area
c) 3,000 square feet (279 square meters) gross floor area
d) 4,000 square feet (372 square meters) gross floor area

84) Which rating system is eligible for a LEED precertification?

a) LEED BD+C: New Construction and Major Renovations
b) LEED BD+C: Core and Shell Development
c) LEED BD+C: Schools
d) LEED ID+C: Commercial Interiors

85) Building Product Selection and Procurement evaluates products against five criteria areas. Which of the following is not one of those five criteria areas?

a) Energy performance
b) Climate health
c) Ecosystem health
d) Circular economy

86) A team pursuing the operable windows option of Resilient Spaces asks which technical criteria the windows must meet. What does the option require?

a) The windows must be motorized with battery backup
b) The windows must total at least 4% of each room's floor area
c) The windows must meet the opening size and location requirements of ASHRAE 62.1-2022, Section 6.4
d) The windows must pass ASTM E783 field air leakage testing

87) Several products are proposed for a project pursuing the low-emitting materials credit. Which of the following is not eligible to be counted in any low-emitting materials product category?

a) Interior gypsum wallboard installed on nonstructural wall framing
b) Suspended acoustic ceiling panels with their suspension grid
c) Poured concrete used as the building's structural slab
d) Permanently installed systems furniture with seating

88) A team pursuing the Reduce Peak Thermal Loads credit is confirming the building envelope compliance obligation that applies to every option. Which statement is correct?

a) Under any option, the envelope must meet ASHRAE 90.1, Section 5.5, Prescriptive Building Envelope Compliance Path, or Section 5.6, Building Envelope Trade-Off Compliance Path, using the version of ASHRAE 90.1 referenced in the Minimum Energy Efficiency prerequisite

b) The envelope must meet ASHRAE 90.1, Section 5.5, Prescriptive Building Envelope Compliance Path only, and always the most recently published edition of the standard

c) The envelope compliance path and the applicable version of ASHRAE 90.1 are selected independently for each option the team pursues

d) The envelope requirement applies only to options that involve energy modeling; the prescriptive options have no envelope compliance obligation

89) A LEED AP is documenting the filtration approach for a central HVAC system that supplies recirculated air to regularly occupied spaces under Fundamental Air Quality. Which of the following does not satisfy the prerequisite's filtration requirement?

a) A minimum efficiency reporting value (MERV) of 8 in accordance with ASHRAE Standard 52.2-2017.

b) A minimum efficiency reporting value (MERV) of 13 in accordance with ASHRAE Standard 52.2-2017.

c) An equivalent filtration media class of ePM1 50% as defined by ISO 16890-2016.

d) In-room air cleaning systems integrated into the building's controls and systems.

90) Under the Reduce Embodied Carbon credit, Option 2, Path 2, how is a qualifying category counted?

a) Each material type that beats its benchmark counts as one category

b) Each supplier that provides an EPD counts as one category

c) Each project phase counts as one category

d) Each impact category counts as one category

91) While planning electricity submetering for a project subject to ASHRAE Standard 90.1, a team finds that exterior lighting represents a very small share of the whole-building electrical load. Under the energy metering and reporting prerequisite, an end use may be combined with another category when it accounts for less than what share of the whole-building electrical load?

a) Less than 10% of the whole-building electrical load

b) Less than 5% of the whole-building electrical load

c) Less than 15% of the whole-building electrical load

d) Less than 25% of the whole-building electrical load

92) Under Enhanced Energy Efficiency, Option 2 (Energy Simulation), a New Construction project demonstrates improvement using the future source energy metric. For electricity in the U.S., which site-to-source energy conversion factor must the project apply?

a) 1.9
b) 2.0
c) 1.09
d) 2.5

93) Under the Compact and Connected Development credit, a project pursuing the walkable location option by counting uses identifies, within walking distance, five restaurants, three banks, two supermarkets, and one pharmacy. After applying the counting restrictions, how many qualifying uses may the project count?

a) 11 uses
b) 8 uses
c) 6 uses
d) 7 uses

94) A mechanical contractor on a project pursuing the Fundamental Refrigerant Management prerequisite is installing a variable refrigerant flow system that uses long, field-assembled refrigerant piping. For this system with field-assembled joints, what must be performed prior to charging with refrigerant?

a) A visual inspection for oil residue only
b) A leak check and a pressure check, but no vacuum check
c) Recovery, recycling, and reclamation of the refrigerant
d) A leak check, a vacuum check, and a pressure check

95) Under the Reduce Embodied Carbon credit, Option 1, what scope of life-cycle analysis is required?

a) Cradle-to-grave, modules A–C
b) Cradle-to-gate, modules A1–A3
c) Cradle-to-site, modules A1–A4
d) Gate-to-grave, modules B–C

96) A team beginning its human impact assessment asks which methods LEED suggests for gathering the required data. Which set is identified?

a) Focus groups limited to future building occupants
b) A single online survey distributed to project team members
c) Site analysis, community outreach, census reports, GIS mapping, and partnerships with local organizations
d) Building energy modeling, life-cycle assessment, and commissioning reports

97) A developer is building on a zero lot line site where the new structure occupies the entire parcel and no exterior site disturbance will occur. The team questions whether a full erosion and sedimentation control plan and site assessment are still mandatory. According to the Minimized Site Disturbance prerequisite, what is the correct approach?

a) Submit a full erosion and sedimentation control plan but omit the site assessment, since no vegetation is present

b) Pay a fee in lieu of the plan because zero lot line projects cannot physically implement erosion controls

c) Apply a reduced inspection schedule of once every 30 calendar days in place of the standard plan

d) Develop a narrative describing why no erosion and sediment control plan or site assessment is necessary for the site

98) A project team proposes pursuing an Innovation Strategy under Project Priorities by adopting a measure not addressed anywhere in the LEED v5 rating system. Which of the following is not one of the criteria the team is required to identify in its submittal documentation?

a) The intent of the proposed innovation strategy

b) A peer-reviewed life-cycle assessment validating the strategy against ISO 14040

c) Proposed requirements for compliance

d) The design approach or strategies used to meet the requirements

99) Under the Air Quality Testing and Monitoring credit, under path 1 of the preoccupancy air testing option, a team measures contaminants over a four-hour period. For which contaminant must the team calculate the peak concentration rather than the average concentration?

a) Ozone

b) PM2.5

c) Carbon monoxide

d) PM10

100) LEED Reference Guides are published by:

a) GBCI

b) USGBC

c) GBCI and USGBC

d) LEED Council

PRACTICE TEST 1 – ANSWERS

A score above 80 would indicate very well preparation for the exam.

1	A	**26**	B	**51**	C	**76**	B
2	B	**27**	B	**52**	D	**77**	D
3	B	**28**	A	**53**	D	**78**	C
4	D	**29**	C	**54**	B	**79**	B
5	A	**30**	B	**55**	A	**80**	A
6	B	**31**	B	**56**	B	**81**	B
7	A	**32**	C	**57**	C	**82**	B
8	B	**33**	D	**58**	A	**83**	A
9	A	**34**	C	**59**	C	**84**	B
10	D	**35**	C	**60**	A	**85**	A
11	D	**36**	C	**61**	B	**86**	C
12	B	**37**	B	**62**	D	**87**	C
13	A	**38**	B	**63**	D	**88**	A
14	B	**39**	A	**64**	B	**89**	A
15	B	**40**	B	**65**	D	**90**	A
16	C	**41**	C	**66**	C	**91**	A
17	A	**42**	C	**67**	C	**92**	B
18	C	**43**	B	**68**	C	**93**	D
19	D	**44**	A	**69**	D	**94**	D
20	D	**45**	C	**70**	C	**95**	A
21	B	**46**	D	**71**	A	**96**	C
22	C	**47**	C	**72**	C	**97**	D
23	B	**48**	B	**73**	C	**98**	B
24	D	**49**	B	**74**	A	**99**	C
25	B	**50**	C	**75**	A	**100**	B

PRACTICE TEST 1 ANSWERS & EXPLANATIONS

1) Correct answer: A. A recycling program.

Every LEED project must have a recycling program. The prerequisite does not mandate on-site composting, a waste-to-energy contract, or a zero-landfill guarantee.

2) Correct answer: B. It is exempt from the prerequisite.

Core and Shell projects without fixtures and fittings or water-using systems are exempt from the prerequisite. Because nothing in the scope uses water, there is no consumption to monitor, so no whole-building meter is required.

3) Correct answer: B. The shared socioeconomic pathways applied.

The prerequisite requires the team to specify the Intergovernmental Panel on Climate Change emissions scenario used, specifying the shared socioeconomic pathways. The shared socioeconomic pathways are the named scenario families that pair a warming trajectory with an assumed socioeconomic future, so identifying the pathway is what makes the scenario meaningful and reviewable.

4) Correct answer: D. Choice in furniture configuration and a variety of seating accommodating a wide range of body types.

Under Occupant Experience, the adaptable environment option's additional strategies include providing choice in furniture configuration and a variety of seating to accommodate a wide range of body types — including seating with back rests and without arm rests. Armless seating serves larger bodies and transfer from mobility devices; backed seating serves those needing support. Uniform furniture is the opposite of the option's premise, which pairs sensory variability with at least one accessible quiet space and one additional strategy.

5) Correct answer: A. The baseline system designated for the building using ASHRAE Standard 90.1, Appendix G must include a cooling tower.

The optimize water use for cooling path has a specific eligibility gate: the baseline system designated using ASHRAE Standard 90.1-2019 or 90.1-2022, Appendix G, must include a cooling tower. The 10%-of-total-water and the 20%-alternative-water conditions belong to the process water use path, not this one. The one-time potable water analysis with five control parameters is the entry requirement for the limit cooling tower cycles path, a sibling path under the same option. Eligible projects then earn 1 point for a 25% reduction and 2 points for a 50% reduction compared to a water-cooled chiller system with axial variable-speed fan cooling towers at three cycles.

6) Correct answer: B. 7,200 kBtu/hr (600 tons or 2,110 kW).

Enhanced Commissioning, Option 2, Path 2 (Enhanced Software) requires fault detection and diagnostics (FDD) for projects with large HVAC or refrigeration capacity—specifically, for total project installed capacity of cooling, heating, or refrigeration systems exceeding 7,200 kBtu/hr (600 tons or 2,110 kW). At that threshold the project must provide a remotely accessible FDD system addressing at least 60% weighted by capacity of air handling equipment and large hydronic or commercial refrigeration equipment.

7) Correct answer: A. ASHRAE 62.1-2022 and 62.2-2022.

Acceptable indoor air quality is achieved per ASHRAE 62.1-2022 and 62.2-2022 — the ventilation standards for commercial and residential spaces respectively. The other pairings are energy, thermal comfort, filtration, or air-cleaning standards.

8) Correct answer: B. It is an acceptable means of diversion and is not considered waste-to-energy for LEED project diversion reporting, unlike all other incineration.

The credit states that combustion of recovered wood materials from recycling processing, known as wood-derived fuel or biomass, is an acceptable means of diversion and is not considered waste-to-energy for LEED diversion reporting. By contrast, all other forms of waste-to-energy or incineration are not considered diversion under any circumstances.

9) Correct answer: A. At least three days.

The power resilience option requires the project to identify critical equipment needing continuous operation and to be able to island and operate independently from the grid, powering those critical loads with the project's on-site renewable and energy storage systems for at least three days (equivalently, at least 72 hours).

10) Correct answer: D. At least three shared-use space types for public use spaces, and at least two dedicated-use space types for community partnership.

The public use spaces path requires making at least three eligible space types (such as auditorium, gymnasium, or cafeteria) accessible for shared public use, while the community partnership path requires contracting for at least two dedicated-use space types (such as a health clinic or library).

11) Correct answer: D. For equipment with refrigerants greater than 700 GWP, and for at least 80% of the total GWP of refrigerants used in the project.

Because field-installed piping leaks far more than self-contained units, the credit mandates self-contained equipment with no field-installed piping for equipment using refrigerants greater than 700 GWP, and for at least 80% of the total GWP of refrigerants used in the project.

12) Correct answer: B. On the project site, on the campus where the project sits, or on the site of a social impact project.

Tier 1 covers on-site renewable generation or a social impact project: the equipment may sit on the project site, on the campus where the project is located, or on the site of a social impact project. This proximity is what makes Tier 1 the highest-value tier. Nationwide siting and off-site certified electricity describe the lower tiers, and Tier 1 is not restricted to the project's own roof — campus and social-impact siting also qualify.

13) Correct answer: A. At least 10 species native or adapted to the project's EPA Level III ecoregion.

Vegetation restoration must include at least 10 species that are native or adapted to the project's EPA Level III ecoregion (or local equivalent for projects outside the U.S.), drawn from a minimum of two of the categories trees, shrubs, and ground cover, with zero lot line projects exempt from the two-category rule.

14) Correct answer: B. Ceilings.

In Building and Materials Reuse, Option 2, the targeted materials are carpeting, ceilings, furniture (ancillary and systems), and interior walls; these are prioritized for high embodied carbon, toxic landfill impacts, and recovery potential, and they receive a 2× multiplier compared to other reused materials. Dimensional lumber, doors, and plumbing fixtures are listed under “other materials,” which do not receive the multiplier. The thresholds are: reuse at least 30% of 1 targeted type, or 15% of 2 targeted types, or 15% of 4 other types, for 2 points.

15) Correct answer: B. An interior door is documented under the walls category.

The walls category includes finish wall treatments, finish carpentry, gypsum wallboard, wall base, interior and exterior doors, and nonstructural wall framing, so a door belongs in the walls category. The other choices misroute products: foamed-in-place and sprayed insulation are expressly excluded from paints and coatings and belong in the insulation category; countertops are excluded from walls and assigned to the furniture category; and wet-applied products applied on the floor are excluded from flooring (wet-applied products fall under paints and coatings or adhesives and sealants depending on type).

16) Correct answer: C. Makeup water must not exceed 5,000 gallons (18,927 liters) of potable water per year, or at least 75% of annual makeup water must come from alternative water sources.

Under Enhanced Resilient Site Design, the drought hazard requires compliance with Enhanced Water Efficiency, native and/or drought-tolerant plantings, and makeup water for created water features that complies with SITES C3.4 or local equivalent. For Core and Shell projects only, makeup water for any created water features must not exceed 5,000 gallons (18,927 liters) of potable water per year, or at least 75% of annual water makeup must come from alternative water sources.

17) Correct answer: A. At least one meeting focused on mechanical, electrical, and plumbing and one focused on the envelope, which may be separate or combined.

Under Fundamental Commissioning, the CxP must attend at least one meeting focused on mechanical, electrical, and plumbing and one focused on the envelope during design, to discuss review comments and commissioning; the meetings may be separate or combined. Enhanced Commissioning raises the bar — at least two design coordination meetings for the MEP path, and for the Core and Shell combined path, two design meetings plus four construction milestone meetings.

18) Correct answer: C. A ratio exceeding 1:2 within 0.5 miles (800 meters) walking distance.

To satisfy this condition the project must be in a community where the jobs-to-housing ratio exceeds 1:2 within 0.5 miles (800 meters) walking distance, meaning more than roughly 4.8 jobs for every four housing units. This jobs-to-housing condition is one of three alternatives, the other two being compliance with the access-to-transit or walkable-location options of the compact and connected development credit.

19) Correct answer: D. Signage in all languages spoken by more than 5% of the local population.

Under Accessibility and Inclusion, the social health category includes providing signage in all languages spoken by more than 5% of the local population. The threshold makes the local demographic data from the Human Impact Assessment directly actionable. Neighboring social health strategies include lactation room pods, a fully accessible all-gender restroom on each floor or a multiuse all-gender restroom, and adult changing stations. Nontext diagrams and symbols are a separate strategy in the navigation category — a complement to multilingual signage, not a substitute.

20) Correct answer: D. ASCE/SEI 7.

For projects in high-wind areas, the credit requires the site and site structures to comply with wind design measures per ASCE/SEI 7 in specified FEMA zones, or a local equivalent. ASCE 24 is for flood-resistant design, not wind. The FORTIFIED Commercial Wind standards are tied specifically to hurricane-prone areas, so they are not the general high-wind reference, and FEMA 543 is cited with ASCE 24 for the flooding hazard.

21) Correct answer: B. A 25-year period during which the grid emissions factor declines in a straight line by 95%.

Under Operational Carbon Projection and Decarbonization Plan, USGBC generates a business-as-usual (BAU) projection of the project's operational carbon emissions over the next 25 years.

22) Correct answer: C. 30%.

On Path 1, which excludes any on-site renewable contribution, the maximum 10 points requires a 30% reduction in the performance index below the performance index target. The 100% figure is the 10-point threshold for Path 2, which includes the on-site renewable contribution and therefore sets thresholds more than three times higher.

23) Correct answer: B. LEED impact areas.

Each credit in LEED makes a contribution to one or more of the three LEED impact areas. Credits that most directly address the prioritized impacts are given the greatest weight—meaning they contain more points. For example, a credit that significantly reduces operational carbon emissions will carry more points than a credit with a narrower environmental benefit.

24) Correct answer: D. One additional strategy: clear width for human passage.

Where no accessibility code is in place, projects must demonstrate exactly three foundational physical accessibility elements: ramps for accessible routes and entrances, clear width for human passage, and a front approach wheelchair accessible section. Since two of the three are already named in the question, only one remains, and it is clear width for human passage.

25) Correct answer: B. Energy and Atmosphere, with five.

In LEED v5 BD+C, Energy and Atmosphere carries five prerequisites: Operational Carbon Projection and Decarbonization Plan, Minimum Energy Efficiency, Fundamental Commissioning, Energy Metering and Reporting, and Fundamental Refrigerant Management. Indoor Environmental Quality carries three, Water Efficiency and Materials and Resources two each, Sustainable Sites one, and the Integrative Process category holds the assessment prerequisites. Location and Transportation carries none.

26) Correct answer: B. Material quantities must not change by more than 5% from design through construction.

Reduce Embodied Carbon, Option 1 requires a cradle-to-grave (modules A–C) WBLCA excluding operating energy and operating water-related energy; baseline and proposed buildings of comparable size, function, orientation, and operating energy performance; and a service life that is the same for both and at least 60 years. The false statement is the 5% figure: the credit allows quantities to use as-designed or as-constructed values provided that quantities did not change more than 10% from design through construction (analysis must be updated if they do).

27) Correct answer: B. Excluded from this credit's diversion calculation; they instead count toward Building and Materials Reuse.

Materials salvaged and reused on-site do not count toward construction and demolition diversion calculations; they instead count toward Building and Materials Reuse. The 200% valuation and inclusion in this credit apply only to materials salvaged and sent off-site for reuse, not to materials kept in place.

28) Correct answer: A. 19 or lower.

Each luminaire may comply with electric light glare control by achieving a unified glare rating (UGR) of 19 or lower, either through the UGR tabular method for each space or through software modeling of the designed lighting performed per the NEMA white paper on unified glare rating. The credit explains that the value of 19 corresponds to just acceptable glare on the psychometric scale of discomfort.

29) Correct answer: C. 100% of parking spaces.

The parking-under-cover option of Heat Island Reduction requires placing 100% of parking spaces under cover. The calculations must include all existing and new off-street parking spaces that are subsidized, leased, or owned by the project, including parking that is outside the project boundary but is used by the project; on-street parking in public rights-of-way is excluded from these calculations.

30) Correct answer: B. Path 1 provides enhanced commissioning for MEP systems, and Path 2 additionally provides building enclosure commissioning.

Option 1 offers Path 1, Enhanced Commissioning for MEP Systems, and Path 2, Enhanced Commissioning for Building Enclosure, which a team can add for further achievement. The two address the building's mechanical/electrical/plumbing systems and its enclosure respectively. Monitoring-based commissioning is Option 2, not a path within Option 1, and the paths are not defined by renewable systems or by the fundamental commissioning scope.

31) Correct answer: B. Option 1 rewards developing a standard green lease and committing to incorporate it into future tenant leases, while Option 2 rewards actually executing that lease with a tenant.

Option 1, Standard Green Lease, rewards writing the document — developing a standard green lease that establishes tenant fit-out standards and incorporates the required tenant requirements, together with a commitment to include it in each future tenant lease. Option 2, Executed Standard Green Lease, rewards signing one with a real tenant, demonstrating the commitment has moved from policy to practice (a signed and dated letter of attestation from the tenant is accepted when the lease is not yet finalized).

32) Correct answer: C. A 30% reduction from the project's baseline.

Under Enhanced Water Efficiency, the whole-project water use option requires demonstrating a minimum 30% reduction from the project's baseline to begin earning recognition, with deeper reductions rewarded on an ascending scale and additional recognition for reductions served by alternative water sources.

33) Correct answer: D. A signed and dated letter of attestation confirming the tenant's commitment to the sustainability criteria in the green lease.

Option 2 awards one point for an executed tenant green lease or, if the tenant has not yet finalized the lease, a signed and dated letter of attestation confirming the tenant's commitment to meet the specific sustainability criteria outlined in the green lease, including compliance with the required prerequisites and best practices.

34) Correct answer: C. No, because New Construction Platinum requires 8 points in the Enhanced Energy Efficiency credit.

LEED v5 Platinum requires both 80+ points and meeting all four decarbonization thresholds for New Construction: 5 points in Electrification, 8 points in Enhanced Energy Efficiency, 100% renewable energy from any combination of Tier 1, 2, and 3, and 20% embodied carbon reduction. With only 6 points in Enhanced Energy Efficiency, the project falls short of the required 8, so it defaults to Gold despite exceeding the point threshold. Tier 3 RECs do count toward the renewable energy requirement.

35) Correct answer: C. The building owner or the owner's representative, who must attest that they have reviewed them.

The building owner or owner's representative must attest that they have reviewed the business-as-usual carbon projection, and the same attestation applies to the decarbonization plan where one is required.

36) Correct answer: C. At least 10%.

The Fundamental Commissioning prerequisite requires the commissioning provider to review at least 10% of the contractor's completed commissioning documents as a quality assurance sample review.

37) Correct answer: B. It is in lighting zones 3 and 4 and automatically turned off from midnight until 6 a.m.

Lighting used solely for facade and landscape lighting qualifies for the exemption only when it is in MLO lighting zones 3 and 4 and is automatically turned off from midnight until 6 a.m., and it must be controlled separately from nonexempt lighting.

38) Correct answer: B. 1% of all regular building occupants.

The Shower and Changing Facilities path requires providing access to on-site showers with changing facilities for 1% of all regular building occupants; off-site facilities are acceptable if they meet occupant needs and lie within 0.25 miles (400 meters) walking distance.

39) Correct answer: A. A third-party verification of the achieved environmental benefit.

For an innovation strategy the team must identify the intent of the proposed innovation strategy, proposed requirements for compliance, proposed submittals to demonstrate compliance, and the design approach or strategies used to meet the requirements. A separate third-party verification of the achieved benefit is not among the listed criteria, making it the correct answer to this negative question; the other three are explicitly required.

40) Correct answer: B. They must be included in the calculations because they are within the LEED project boundary.

The prerequisite states that all ancillary structures, such as parking structures or outbuildings within the LEED project boundary, must be included in the calculations, and the materials-to-include guidance specifically lists parking structures and parking lots. Whether a structure is attached or detached, occupied or unoccupied, is irrelevant; the controlling test is whether it lies within the LEED project boundary.

41) Correct answer: C. Computers, kitchen refrigerators and stoves, washers and dryers, and elevators and escalators.

Process energy, also called unregulated energy, is tied to building use and operations rather than occupant comfort — computers and office equipment, kitchen stoves and refrigerators, cooking and food-preparation equipment, washers and dryers, and elevators and escalators. These are precisely the plug, process, refrigeration, and conveyance loads addressed by the credit's plug-and-process-loads path. Lighting, HVAC, hot water heating, parking ventilation, and toilet exhaust are all regulated (nonprocess) loads.

42) Correct answer: C. 7 spaces, including at least one in an accessible parking space.

For commercial projects, EVSE must be installed for at least 5% of total parking capacity and no fewer than two spaces; if the calculation does not yield a whole number, the team rounds up (not down). Here, 125 x 0.05 = 6.25, which rounds up to 7 total required spaces. Critically, at least one of those charging stations must be located in an accessible parking space (at least 9 feet / 2.5 meters wide with a 5-foot / 1.5-meter access aisle and accessibility features).

43) Correct answer: B. Maintain proper filtration on the equipment while it is in use during construction.

The HVAC protection criterion directs teams to keep contaminants out of the HVAC system by avoiding use of permanently installed equipment if possible, but if it must be used, to maintain proper filtration during that use.

44) Correct answer: A. It provides on-site electric or thermal storage with automatic load management controls that store energy during off-peak periods or periods of low grid carbon intensity and deploy it during on-peak periods or periods of high grid carbon intensity.

Option 1 requires on-site electric and/or thermal storage with automatic load management controls that store energy during off-peak periods, or periods of low grid carbon intensity, and deploy it during on-peak periods, or periods of high grid carbon intensity. Shifting consumption away from carbon-intensive, high-demand hours is exactly the grid benefit the credit rewards. The option accepts both electric and thermal storage, is driven by peak and carbon-intensity signals rather than a fixed schedule or emergency-only trigger, and is about time-shifting the project's own use, not exporting surplus.

45) Correct answer: C. At least 30% reduction from the calculated baseline for the site's annual theoretical irrigation requirement (TIR).

The prerequisite's Option 2. Efficient Irrigation requires reducing the project's irrigation water requirement by at least 30% from the calculated baseline for the site's annual theoretical irrigation requirement (TIR), achieved through plant species selection and irrigation system efficiency using the EPA TIR methodology.

46) Correct answer: D. It is not a renewable energy resource.

Geoexchange systems such as geothermal heat pumps that use vapor compression cycles are not considered a renewable energy resource; these systems are credited in Enhanced Energy Efficiency. Eligible Tier 1 resource types are limited to solar electric, solar thermal, wind, and recovered heat from municipal wastewater, so the ground-loop location and the off-site Tier 3 framing are both wrong.

47) Correct answer: C. Option 2 requires retaining on-site the increase in runoff volume relative to pre-disturbance natural land cover.

Under Rainwater Management, Option 2: Natural Land Cover Conditions requires the team to calculate the difference between the projected runoff volume under proposed design conditions and the runoff volume under natural land cover conditions that existed prior to any disturbance, then retain that increase on-site using LID and GI practices. Specific percentile events are not required for Option 2, and it uses a 10-year range of hydrologic rainfall events (or an average representative year), while Option 1 requires a minimum 30-year period of daily rainfall data. The additional water-reuse point is a feature of Option 1, not Option 2, so the other choices are incorrect.

48) Correct answer: B. In lighting zones 2, 3, or 4.

Lighting for the national flag is exempt from the uplight and light trespass requirements in lighting zones 2, 3, or 4, and, like every exemption, only when controlled separately from the nonexempt lighting.

49) Correct answer: B. The LSC and Board of Directors review and affirm, then the consensus committee votes, then USGBC membership ratifies.

LEED improvements are developed by volunteer committees in conjunction with USGBC staff. The proposals are then reviewed and affirmed by the LEED Steering Committee and the USGBC Board of Directors before going to a vote by the relevant consensus committee and ratification by USGBC's membership. This multi-step process ensures transparency, technical rigor, and broad stakeholder support. GBCI administers certification, not rating system development. No single body can publish changes unilaterally.

50) Correct answer: C. Only Option 2 (Enhanced Indoor Air Quality Design) is available; PM2.5 design limit of 10 μg/m3.

Under Enhanced Air Quality, the available compliance path depends on how the project complied with Fundamental Air Quality: if using the VRP, Option 1 or Option 2 may be used, but if using the IAQP, the project must use Option 2 (Enhanced Indoor Air Quality Design). Option 2 applies more stringent design limits beyond ASHRAE 62.1-2022: PM2.5 of 10 μg/m3, formaldehyde of 20 μg/m3, and ozone of 10 ppb. The 12 μg/m3 and 35 μg/m3 figures are preoccupancy testing limits for attainment and nonattainment areas under Air Quality Testing and Monitoring, not the Enhanced Air Quality design target, and 20 μg/m3 is the Enhanced Air Quality formaldehyde limit, not PM2.5.

51) Correct answer: C. At least one fully accessible, all-gender, single-use restroom, or one multiuse all-gender restroom, on each floor of the building.

Under Accessibility and Inclusion, the social health category includes providing at least one fully accessible, all-gender, single-use restroom, or one multiuse, all-gender restroom, on each floor of the building. The per-floor coverage is the substance — a token lobby facility fails users everywhere else in the building. The strategy sits with lactation room pods, adult changing stations, multilingual signage above the 5% threshold, and the biophilic cross-reference for neurodivergent support.

52) Correct answer: D. Replace all air filtration media after construction is complete, then perform testing and balancing with the new filters.

The prerequisite requires teams to replace all air filtration media after completion of construction and before occupancy, and then to confirm that testing and balance work is completed with the new filtration in place. Performing TAB first and replacing media afterward would invalidate the balance, because filter resistance affects airflow. Retaining construction-phase filters, or replacing media only conditionally, both fail the explicit requirement. The correct order is to install the new filters first, then complete testing and balancing with that new filtration.

53) Correct answer: D. Climate zones 3 and above.

The space-heating path states that projects in climate zones 3 and above must have a weighted average space heating equipment efficiency of at least 1.8 coefficient of performance. Space heating equipment in climate zones 0–2 may be excluded from the COP determination entirely, because electric resistance heating contributes far less to peak grid load in hot climates, so it is not the zone that triggers the requirement.

54) Correct answer: B. No once-through cooling with potable water is allowed for any equipment or appliances that reject heat.

The processes standard requires no once-through cooling with potable water for any equipment or appliances that reject heat. Capturing discharge, adding a meter, or invoking emergency backup does not create an exception to that prohibition.

55) Correct answer: A. The guardrail glass must still comply, because glass in guardrails and windshields is regulated at any distance from grade or roof.

Under Biodiverse Habitat, exterior glass must have a maximum threat factor of 30 per the American Bird Conservancy's Threat Factor Database. The height scope covers all glass, including spandrel glass, from grade up to 50 feet (15 meters). It extends up to 20 feet (6 meters) above the finished grade of a green roof. Glass in guardrails and windshields is covered at any distance from grade or roof. The 50-foot ceiling therefore never shields guardrail glass. Exceeding threat factor 30 is a failure, not an exemption, and the green-roof provision extends coverage rather than limiting it.

56) Correct answer: B. Higher-education classroom students are visitors; K–12 students are regular building occupants.

Under LEED v5's project occupancy guidance, regular building occupants are routine full- and part-time users. They include employees, daily volunteers, support staff, residents, overnight hotel guests, medical inpatients, and K–12 students. Visitors are intermittent users: retail customers, medical outpatients, periodic volunteers, and higher-education classroom students. College students attend intermittently; K–12 students attend daily. When occupancy cannot reasonably be projected, teams use ASHRAE 62.1-2022 default densities or the reference guide's table, applied as both daily average and peak.

57) Correct answer: C. The owner's project requirements.

The credit's goal-setting initiative directs the team to define specific and measurable project goals addressing decarbonization, quality of life, and ecological conservation and restoration, and to incorporate these goals into the owner's project requirements. The basis of design describes how the design responds to those requirements and is prepared by the design team, so it is downstream of the OPR rather than the document the credit names.

58) Correct answer: A. 80%.

Under the submeters option, the indoor plumbing fixtures and fittings subsystem requires metering systems serving at least 80% of indoor fixtures and fittings as described in the minimum water efficiency prerequisite. Fixtures not addressed in the prerequisite, such as janitor sinks and bottle fillers, may be included or excluded at the team's discretion.

59) Correct answer: C. at least 75% of the occupied spaces.

The acoustic criteria path requires demonstrating, through calculations, modeling, and/or measurements, that the design meets acoustic criteria for at least 75% of the occupied spaces and all classrooms and other core learning spaces. For a school, every classroom and core learning space must comply because acoustics strongly affect learning and teacher well-being, while the broader building must reach the 75% threshold.

60) Correct answer: A. The existing building structure and enclosure.

Option 1 rewards maintaining the existing building structure and enclosure.

61) Correct answer: B. It counts; the option addresses reuse of structure and enclosure materials left in situ as well as those procured off-site and incorporated into the building's structure or enclosure.

Under Building and Materials Reuse, the building reuse option addresses on-site reuse of existing structure and enclosure materials, including materials left in situ and those procured off-site and incorporated into the building's structure or enclosure. Salvaged structural steel bought from another site and built into the frame counts toward the reuse percentage. Teams should engage architects and structural engineers from early design to identify reusable structural elements. Off-site salvage counts under waste diversion only when this project ships materials out — inbound procured salvage belongs here.

62) Correct answer: D. Level 3, with a 3x multiplier.

The credit describes Level 3 products as elite and representing the forefront of sustainability, and products scored at this level earn a 3x multiplier. Level 1 earns a 1x multiplier and Level 2 earns a 2x multiplier; there is no 5x multiplier in the scoring.

63) Correct answer: D. All entries, outdoor air intakes, and operable windows.

No Smoking or Vehicle Idling requires that outdoor smoking be prohibited except in designated smoking areas located at least 25 feet (7.5 meters) (or the maximum extent allowable by local codes) from all entries, outdoor air intakes, and operable windows, to minimize the likelihood of smoke entering the building.

64) Correct answer: B. 600 feet (180 meters).

The Bicycle Network and Storage path requires that a functional entry and/or bicycle storage be within a 600-foot (180-meter) walking or bicycling distance of a qualifying bicycle network. The distractor of 300 feet (90 meters) is the threshold for long-term bicycle storage placement from a functional entry, not the network distance. The 0.25-mile (400-meter) figure governs acceptable off-site shower and changing facilities, while 0.5 miles (800 meters) is the transit-stop walking distance used in the assessment exemption.

65) Correct answer: D. ASHRAE 90.1-2019 with Addendum cr.

Option 1 requires compliance with ANSI/ASHRAE/IES Standard 90.1-2019 with Addendum cr, which mandates a building envelope backstop for projects documented using the Energy Cost Budget Method or Appendix G Performance Rating Method; all other 90.1-2019 addenda are optional.

66) Correct answer: C. Meters sufficient to capture total potable water use, with a minimum of one per floor.

For Core and Shell, the submeters option requires that meters be installed for future tenant spaces so tenants can independently meter their water consumption, providing enough meters to capture total potable water use with a minimum of one per floor.

67) Correct answer: C. Option 2. Management Mode for Respiratory Diseases.

Under Resilient Spaces, both Option 1 (Management Mode for Episodic Outdoor Ambient Conditions) and Option 2 (Management Mode for Respiratory Diseases) are designated New Construction Only. The options available to Core and Shell are Option 3 (Design for Occupant Thermal Safety During Power Outages, via Path 1 Extreme Heat and/or Path 2 Extreme Cold) and Option 4 (Operable Windows).

68) Correct answer: C. Between 90% and 110%.

The balanced ventilation requirement directs the design to keep total supply ventilation airflow within 10% of the total exhaust airflow, and during construction the testing, adjusting, and balancing report must verify a ratio of ventilation to exhaust airflow between 90% and 110% on a whole-building basis.

69) Correct answer: D. Observed, projected, and future natural hazards that could affect the project site and building function.

Under Climate Resilience Assessment, the team must identify observed, projected, and future natural hazards that could potentially affect the project site and building function. The list of site-specific hazards includes drought, extreme heat, extreme cold, flooding, hurricanes and high winds, hail, landslides, sea level rise and storm surge, tornadoes, tsunamis, wildfires and smoke, and winter storms.

70) Correct answer: C. 40% of the gross roof area.

The solar readiness criteria require the project to designate a dedicated solar zone area equal to at least 40% of the gross roof area.

71) Correct answer: A. It must be greater than or equal to the sum of the total site paved area and the total roof area.

Option 1 requires teams to use any combination of nonroof, high-reflectance roof, and vegetated roof strategies so that the weighted sum of site design strategies is greater than or equal to the sum of the total pavement and roof areas. The denominator of the calculation is therefore total site paved area plus total roof area — not paving alone, not roof alone, and not the entire project site.

72) Correct answer: C. More than 80% of each additional product category, earning 2 points total.

Under Low-Emitting Materials for New Construction, Path 2 requires achieving Path 1 plus any two of the additional categories (adhesives and sealants, walls, insulation, composite wood) at greater than 80% of each additional product category, which brings the credit to 2 points total.

73) Correct answer: C. Entirely outside any floodplain subject to a 1% or greater chance of flooding in any given year.

Under Sensitive Land Protection, a legally adopted flood hazard map governs where one exists. Where none exists, the development footprint must sit entirely outside any floodplain with a 1% or greater annual chance of flooding. This is the 100-year floodplain.

74) Correct answer: A. 25 feet (7.5 meters).

The view content — nature, urban landmarks, art, or objects — must be at least 25 feet (7.5 meters) from the exterior of the glazing, providing depth of field.

75) Correct answer: A. The LEED prerequisites and credits pursued by the base building.

The guide directs teams to refer to the attempted prerequisites and credits of the base building for content requirements and states that the strategies offered to tenants must align with the LEED prerequisites and credits pursued by the base building.

76) Correct answer: B. Biophilic design requires repeated and sustained engagement with nature.

Under Occupant Experience, the integrated biophilic design path demonstrates five principles adapted from The Practice of Biophilic Design by Kellert and Calabrese. Biophilic design requires repeated and sustained engagement with nature. It focuses on human adaptations to the natural world that have advanced health, fitness, and well-being over evolutionary time. It encourages an emotional attachment to the building and its location. It promotes positive interactions between people and nature that expand relationship and responsibility. And it encourages mutually reinforcing, interconnected, and integrated architectural solutions. The principles are design philosophy, not prescriptive features or metrics.

77) Correct answer: D. Encourage LEED certification for all new construction projects worldwide.

USGBC's six strategic goals are: Sustainable Cities and Communities, Climate and Natural Resources, Green Building Marketplace, Public Policy, International, and Organizational Excellence.

78) Correct answer: C. No more than two uses in each type of use may be counted.

The Walkable Location restrictions state that no more than two uses in each type of use may be counted — for example, if five restaurants are within walking distance, only two may be counted toward Table 4. A use may be counted under only one use type, not several, so the choice allowing a use to count under multiple types is false. The category rule is that the counted uses must represent at least three of the five categories, not four. The counting distance is 0.5 miles (800 meters) from a functional entry, and there is no rule excluding uses within 0.25 miles (400 meters).

79) Correct answer: B. Space for organics/food waste recycling must still be provided even though service is not yet available.

Under Planning for Zero Waste Operations, the recyclable materials handled must include organics/food waste, mixed paper, corrugated cardboard, glass, plastics, and metals, and the prerequisite is explicit that space for the storage of organics/food waste recycling is required even if service is not available at the time of building occupancy. This is driven by the substantial greenhouse gas emissions (methane) from discarded organic waste in landfills.

80) Correct answer: A. No; projects that do not own or lease parking are not eligible for the credit.

Under Electric Vehicles, projects that do not own or lease parking are not eligible for the credit. Both options act on the project's own parking supply — installing EVSE in a share of total spaces, or making spaces EV-ready with full-circuit infrastructure — and a project without controlled parking has nothing to act on.

81) Correct answer: B. Quantified and disclosed embodied carbon impacts.

For Level 1 in climate health, building products must have quantified and disclosed embodied carbon impacts, prioritizing the quantification and disclosure of life-cycle global warming potential through product-specific Type III EPDs. Demonstrating a reduction in carbon via an EPD comparative assessment is reserved for Levels 2 and 3. Disclosing product ingredients describes Level 1 of human health, and incorporating recycled or bio-based content describes Level 1 of circular economy, making both plausible cross-area distractors.

82) Correct answer: B. Weighted average across all new equipment.

The comparison is a weighted average of refrigerant GWP across all new refrigerant-containing equipment, not a single-unit test — one small high-GWP unit will not fail a project whose overall average stays under the threshold, and vice versa.

83) Correct answer: A. 1,000 square feet (93 square meters) gross floor area.

According to LEED Minimum Program Requirements (MPRs), projects must comply with specific size thresholds to be eligible for certification. LEED BD+C projects must include a minimum of 1,000 square feet (93 square meters) of gross floor area.

In addition to meeting size requirements, all LEED projects must satisfy the other two Minimum Program Requirements: the project must be in a permanent location on existing land and must use reasonable LEED boundaries.

84) Correct answer: B. LEED BD+C: Core and Shell Development.

LEED BD+C: Core and Shell Development projects are eligible for precertification.

85) Correct answer: A. Energy performance.

Under Building Product Selection and Procurement, nonstructural products are evaluated against five criteria areas: climate health, human health, ecosystem health, social health and equity, and circular economy. Energy performance is not one of them—that belongs to the Energy and Atmosphere category. These five areas align with the Mindful Materials Common Materials Framework and the AIA A&D Materials Pledge. Products achieving two or more criteria areas are considered multi-attribute and earn higher value in the credit calculation. Exam tip: memorize the five as climate, human, ecosystem, social (health and equity), and circular economy—no "energy" category.

86) Correct answer: C. The windows must meet the opening size and location requirements of ASHRAE 62.1-2022, Section 6.4.

Under Resilient Spaces, the operable windows option designs regularly occupied spaces with operable windows capable of providing access to outdoor air during heat waves or localized power outages. The windows must meet the opening size and location requirements of ASHRAE 62.1-2022, Section 6.4 — the natural ventilation provisions. Section 6.4 governs openable area and the geometry between openings and occupied zones, ensuring the windows actually deliver air, not just daylight. Motorization and leakage testing are not the option's criteria.

87) Correct answer: C. Poured concrete used as the building's structural slab.

The credit explicitly lists products that are not applicable to the low-emitting materials categories, including fire suppression, HVAC and ductwork, plumbing, electrical, conveying and communications systems, poured concrete, structural framing, structural insulated panels, and water-resistive barriers. Gypsum wallboard is expressly included in the walls category, ceiling panels and suspension grids are included in the ceilings category, and systems furniture and seating are included in the furniture category.

88) Correct answer: A. Under any option, the envelope must meet ASHRAE 90.1, Section 5.5, Prescriptive Building Envelope Compliance Path, or Section 5.6, Building Envelope Trade-Off Compliance Path, using the version of ASHRAE 90.1 referenced in the Minimum Energy Efficiency prerequisite.

The requirement has three parts that this question ties together. First, it applies to all options of the Reduce Peak Thermal Loads credit — no option is exempt. Second, the envelope may comply through either ASHRAE 90.1, Section 5.5, Prescriptive Building Envelope Compliance Path, or Section 5.6, Building Envelope Trade-Off Compliance Path — both are acceptable, so it is not restricted to the prescriptive route. Third, the applicable edition is the version of ASHRAE 90.1 referenced in the Minimum Energy Efficiency prerequisite, which keeps the project on one consistent edition rather than the newest published one or a version chosen per option.

89) Correct answer: A. A minimum efficiency reporting value (MERV) of 8 in accordance with ASHRAE Standard 52.2-2017.

Under Fundamental Air Quality, each central HVAC system supplying outdoor and/or recirculated air to regularly occupied spaces must meet one of several options, including MERV 13 per ASHRAE Standard 52.2-2017, an equivalent ePM1 50% media class per ISO 16890-2016, in-room air cleaning systems, or systems tested for effectiveness and safety per ASHRAE Standard 241-2023. MERV 8 does not meet the requirement.

90) Correct answer: A. Each material type that beats its benchmark counts as one category.

Path 2, the Materials-Type Approach, credits beating the industry benchmark within individual material categories — each material type that clears its benchmark counts as one qualifying category.

91) Correct answer: A. Less than 10% of the whole-building electrical load.

Teams may combine electricity end uses that are less than 10% of the whole-building electrical load with other categories; for example, exterior lighting under 10% of the whole-building load may be reported with interior lighting.

92) Correct answer: B. 2.0.

Enhanced Energy Efficiency, Option 2 requires improvement in future source energy, and for U.S. projects electricity is converted using a national average electric site-to-source energy conversion factor of 2.0, based on projections through 2050.

93) Correct answer: D. 7 uses.

The restrictions cap each use type at no more than two uses. The five restaurants count as only 2, the three banks count as only 2, the two supermarkets count as 2, and the one pharmacy counts as 1, totaling 7 qualifying uses. The team must also confirm the counted uses represent at least three of the five categories; here food retail, services, and community-serving retail are represented, so that condition is met.

94) Correct answer: D. A leak check, a vacuum check, and a pressure check.

The prerequisite requires that, prior to substantial completion, both new and existing refrigerant-containing equipment be checked for leaks and that all identified leaks be repaired. Critically, for systems with field-assembled joints, the team must perform a leak check, a vacuum check, and a pressure check prior to charging with refrigerant. A visual inspection alone is acceptable only for existing or self-contained systems, not field-assembled piping.

95) Correct answer: A. Cradle-to-grave, modules A–C.

Option 1 requires a cradle-to-grave whole-building life-cycle assessment (modules A–C) of the structure, enclosure, and hardscape, excluding operating energy and water.

96) Correct answer: C. Site analysis, community outreach, census reports, GIS mapping, and partnerships with local organizations.

Under Human Impact Assessment, teams should use methods such as site analysis, community outreach, census reports, GIS mapping, and partnerships with local organizations. These sources populate the assessment's categories — demographics, local infrastructure and land use, human use and health impacts, and occupant experience — and help validate data against local knowledge.

97) Correct answer: D. Develop a narrative describing why no erosion and sediment control plan or site assessment is necessary for the site.

The prerequisite states that zero lot line projects and projects that cause no exterior site disturbance should develop a narrative that describes why no erosion and sediment control (ESC) plan or site assessment is necessary for the site.

98) Correct answer: B. A peer-reviewed life-cycle assessment validating the strategy against ISO 14040.

For an Innovation Strategy not addressed in the LEED v5 rating system, the team must demonstrate significant, measurable environmental performance and identify four items: the intent of the proposed innovation strategy, the proposed requirements for compliance, the proposed submittals to demonstrate compliance, and the design approach or strategies used to meet the requirements.

99) Correct answer: C. Carbon monoxide.

Path 1 requires measuring for a four-hour period and calculating the peak concentration for carbon monoxide and the average concentration for ozone, PM2.5, and PM10.

100) Correct answer: B. USGBC.

LEED Reference Guides are published by USGBC.

SECTION 3:

PRACTICE TEST 2

1) What distinguishes a LEED interpretation from other types of LEED guidance?

a) It applies only to the project that requested it
b) It is a suggestion that projects may ignore
c) It is precedent-setting and applies to all LEED projects
d) It expires after one year

2) A team relies on a qualified independent laboratory report (not a third-party certification) to demonstrate VOC compliance for a flooring product under the low-emitting materials credit. Which condition regarding the timing of the laboratory test must be satisfied?

a) The product must have been tested no more than three years prior to the product's purchase
b) The product must have been tested no more than one year prior to the product's purchase
c) The product must have been tested no more than two years prior to the product's purchase
d) The product must have been tested no more than five years prior to the product's purchase

3) Under the erosion and sedimentation control plan required by Minimized Site Disturbance, a project team elects to inspect controls every 14 calendar days rather than every seven. Which additional inspection trigger must the plan also satisfy to remain compliant?

a) An inspection within 48 hours of any storm event producing 0.5 inches (13 millimeters) or more of rain in a 24-hour period.

b) A weekly inspection of dewatering discharges only, with no storm-triggered requirement.

c) An inspection within 24 hours of the occurrence of a storm event that produces 0.25 inches (6 millimeters) or more of rain within a 24-hour period.

d) An inspection within 72 hours of any storm event, plus monthly inspections of construction-exclusion zones.

4) Planning for Zero Waste Operations directs project teams to follow the true waste hierarchy framework, which ranks waste management strategies from most to least preferred. Which sequence correctly reflects that order?

a) Recycle, Reduce, Reuse, Compost, Anaerobic Digestion

b) Reuse, Reduce, Compost, Recycle, Anaerobic Digestion

c) Reduce, Recycle, Reuse, Anaerobic Digestion, Compost

d) Reduce, Reuse, Recycle, Compost, Anaerobic Digestion

5) Under Option 1 of the Green Leases credit, the standard green lease must include tenant requirements ensuring compliance with a defined list of prerequisites for any tenant-installed systems. Which of the following prerequisites is not on that required list?

a) Climate Resilience Assessment

b) Fundamental Commissioning

c) Planning for Zero Waste Operations

d) Building-Level Water Metering

6) Under the Building and Materials Reuse credit, Option 1, which elements are counted?

a) Floor decking, roof decking, and enclosure

b) Windows, doors, and partitions

c) Cladding and roofing only

d) Foundations and footings only

7) A LEED BD+C: Core and Shell project pursues the no-on-site-combustion path of the Electrification credit but, because of its limited base-building scope, addresses only a portion of the combined space heating and service water heating load. Which weighted average coefficient of performance thresholds apply to this Core and Shell path?

a) A higher threshold of 2.0 COP and a lower threshold of 1.5 COP

b) A higher threshold of 1.8 COP and a lower threshold of 1.3 COP

c) A higher threshold of 1.5 COP and a lower threshold of 1.0 COP

d) A single threshold of 1.8 COP, with no lower alternative available to Core and Shell projects

8) For a residential project under the Bicycle Network and Storage path of the Transportation Demand Management credit, the long-term bicycle storage requirement is at least 15% of all regular building occupants but no fewer than:

a) one storage space per two dwelling units
b) one storage space per four dwelling units
c) four storage spaces per building
d) one storage space per three dwelling units

9) A team pursuing the Operable Windows option of the Resilient Spaces credit wants to earn the maximum points available for that option. For what minimum percentage of the regularly occupied spaces must operable windows be designed to provide access to outdoor air?

a) 50%
b) 75%
c) 65%
d) 90%

10) A residential New Construction project with 150 total parking spaces wants the single point offered by the electric vehicle readiness option of the Electric Vehicles credit. What is the minimum number of EV-ready spaces required?

a) 15 spaces
b) 30 spaces
c) 20 spaces
d) 23 spaces

11) A school project pursues Compact and Connected Development credit, Option 4, Path 2: Connected Site. The team wants to qualify the location as an infill site to earn 2 points. What minimum percentage of the project boundary must border parcels that are previously developed sites?

a) At least 75%
b) At least 25%
c) At least 50%
d) At least 100%

12) The carbon projection produced under the Operational Carbon Projection and Decarbonization Plan prerequisite is intended to reveal a specific pattern over time. What does it show?

a) Emissions from electricity and from on-site combustion both decline at the same rate
b) Emissions from electricity decline toward zero over time, while emissions from on-site combustion remain constant
c) Emissions from on-site combustion decline as equipment efficiency improves, while grid emissions remain constant
d) Both sources remain constant, since the projection assumes no change in operations

13) Under Minimum Energy Efficiency, a project may comply by using either Option 1 (ASHRAE 90.1-2019) or Option 2 (ASHRAE 90.1-2022). Which statement about the version requirement is correct?

a) All projects must use ASHRAE 90.1-2022 regardless of registration date

b) Only projects registered before January 1, 2028, may use Option 1; Option 2 is required for projects registered on or after that date

c) Projects registering before January 1, 2026, may use either option; after that date only Option 1 is allowed

d) Projects may use ASHRAE 90.1-2019 only if they register on or after January 1, 2028

14) A team plans to earn points under the exemplary performance pathway for an existing LEED v5 credit. According to the Project Priorities credit, how does exemplary performance earn its points?

a) By exceeding the credit requirements or achieving the next incremental percentage threshold

b) By achieving the credit at the highest available point threshold

c) By documenting an innovative strategy not addressed in the rating system

d) By demonstrating regional importance for the project's location

15) Energy Metering and Reporting requires projects to commit to reporting energy data to USGBC. For how long must this reporting commitment carry forward, and what data must be reported?

a) Three years (or until the building changes ownership or lessee); monthly energy consumption for each source, on-site renewable energy generation, and peak electrical demand

b) Ten years; monthly energy consumption by source and 15-minute interval submetered data

c) Five years; peak electrical demand only

d) Five years (or until the building changes ownership or lessee); monthly energy consumption for each source, on-site renewable energy generation, and peak electrical demand

16) A team is documenting an office building under the prescriptive path of the minimum water efficiency prerequisite. The mechanical engineer must specify standard single-flush water closets that will not exceed the maximum installed flush rate. Which flush rate may the installed toilets not exceed?

a) 1.6 gallons per flush (6.0 liters per flush)

b) 1.0 gallon per flush (3.8 liters per flush)

c) 1.28 gallons per flush (4.8 liters per flush)

d) 2.5 gallons per flush (9.5 liters per flush)

17) For the Design Analysis component of the Operational Carbon Projection and Decarbonization Plan prerequisite, a small, less complex New Construction project wants to minimize documentation burden. Which statement is true regarding its early-design analysis obligations?

a) Every project must develop a calibrated whole-building energy model

b) Only projects pursuing the energy efficiency credit may skip the energy model

c) An energy model is required, but a single-zone box model is never acceptable

d) Smaller, less complex projects are not required to have an energy model

18) A commercial project provides long-term bicycle storage to pursue the Bicycle Network and Storage path of the Transportation Demand Management credit. The storage must be within what walking distance from any functional entry, and at least how many spaces per building?

a) 300 feet (90 meters); no fewer than four spaces

b) 600 feet (180 meters); no fewer than four spaces

c) 300 feet (90 meters); no fewer than two spaces

d) 600 feet (180 meters); no fewer than two spaces

19) For building envelope commissioning under the Fundamental Commissioning prerequisite, which approaches satisfy the requirement?

a) A whole-building air leakage pressurization test, or a continuous air barrier design and installation verification program

b) A whole-building air leakage pressurization test and seismic test

c) Infrared thermographic imaging of the completed enclosure, or a blower door test of a representative floor

d) A manufacturer's warranty certification for the installed air barrier assembly

20) The materials reuse option of Building and Materials Reuse values certain targeted materials more highly than others. Why are the targeted materials singled out?

a) They are the least expensive materials to salvage

b) They are the lightest materials to transport between sites

c) They are exempt from building code approval when reused

d) They have high impacts such as embodied carbon or pollution, are hard to recycle, and significant amounts of them end up in landfill

21) A team pursuing the Heat Island Reduction credit, Option 1, has a shaded, high-reflectance paved area that qualifies under more than one nonroof strategy. How may that area be counted?

a) Once for each qualifying strategy, since the strategies are cumulative

b) Only once, even though it is addressed through multiple strategies

c) Twice at most, provided one strategy is vegetative and the other is material-based

d) Once, but at an increased weighting factor reflecting the multiple strategies applied

22) The required construction management practices in this prerequisite are primarily adapted from a single industry guideline for construction-phase indoor air quality. Which referenced standard is that source?

a) ASHRAE Standard 62.1-2022, Sections 5 and 6, on ventilation for acceptable indoor air quality

b) SMACNA IAQ Guidelines for Occupied Buildings under Construction, 2nd edition, 2007, ANSI/SMACNA 008-2008, Chapter 3

c) ANSI/SMACNA HVAC Duct Construction Standards, Metal and Flexible, 4th edition

d) ASHRAE Standard 52.2-2017 method of testing general ventilation air-cleaning devices

23) A K-12 school project is pursuing the No Smoking or Vehicle Idling prerequisite. Compared with a typical commercial project, which additional smoking requirement applies specifically to the school, and where must its required signage be posted?

a) Prohibit all smoking on-site, with signage posted on the property line

b) Prohibit smoking within 25 feet (7.5 meters) of openings, with signage at each building entrance

c) Prohibit smoking indoors only, with signage posted at designated outdoor smoking areas

d) Prohibit all smoking within 50 feet (15 meters) of the building, with signage at the parking area

24) The five criteria areas of Building Product Selection and Procurement are designed to align with existing industry frameworks. Which pair does LEED identify?

a) The Living Building Challenge Red List and the WELL Building Standard

b) The Mindful Materials Common Materials Framework and the AIA Architecture and Design Materials Pledge

c) ISO 14025 and the EU Ecolabel

d) Cradle to Cradle Certified and Declare

25) A team performing the whole-building life-cycle assessment option incorporates salvaged steel framing into an assembly. The team zeroes out the manufacturing-stage impacts of this reused steel. Regarding the transport-to-site stage (A4) and the use and end-of-life stages (B through C) for that salvaged steel, what does the Reduce Embodied Carbon credit require?

a) Those stages are also zeroed out, because the material is treated as carbon-free once reused

b) Those stages must be included in the analysis as appropriate, even though A1 through A3 are zeroed

c) Only the end-of-life stages may be excluded; the transport stage must be doubled

d) Those stages are optional and may be excluded at the team's discretion

26) A project intends to earn a point under the appliance and process water option of the Enhanced Water Efficiency credit by documenting compliant commercial washing machines. What laundry-throughput threshold must the project meet to be eligible to count these commercial washing machines?

a) The project must process at least 100,000 lb (45,359 kg) of laundry per year

b) The project must include on-premise washers with a minimum capacity of 2,400 lb (10,886 kg) per year

c) The project must process at least 120,000 lb (57,606 kg) of laundry per year

d) The project must serve at least 100 loads of laundry per day of operation

27) A project addressing extreme heat under enhanced resilient site design wants to credit proximity to an off-site community cooling resource for its vulnerable occupants. Within what distance of the building must an emergency cooling station be located to qualify as one of the extreme heat elements?

a) 0.25 miles (400 meters)
b) 0.10 miles (160 meters)
c) 0.50 miles (800 meters)
d) 1 mile (1,600 meters)

28) A team using the ventilation rate procedure to satisfy the fundamental air quality prerequisite pursues the increased ventilation option of the enhanced air quality credit. They model a single-zone, 100% outdoor air system. By what factor must the calculated minimum outdoor airflow for the system be multiplied to determine the increased ventilation requirement for the single point?

a) 1.10
b) 1.30
c) 1.50
d) 1.15

29) A project team pursuing Grid Interactive, Option 3 (Automated Demand-Side Management) proposes using its on-site solar PV generation and a small natural-gas generator to satisfy the demand-side management criteria. A LEED AP advises this approach will fail. Why?

a) On-site electricity generation and fuel combustion cannot be used to meet the demand-side management criteria
b) Option 3 requires energy storage of at least 0.2 kWh/kW relative to peak electric demand
c) Option 3 is only available to Core and Shell projects
d) Demand-side management requires islanding capability for at least three days

30) A project pursuing the Light Pollution Reduction credit has a property line that abuts a public street. How may the team treat the lighting boundary along that edge?

a) The lighting boundary must remain on the property line in all cases
b) The lighting boundary may move to the far curb of the street
c) The lighting boundary may move to the center line of that street
d) The lighting boundary may move 5 feet (1.5 meters) beyond the property line

31) A project team wants to use an alternative standard in place of one referenced by a LEED v5 credit, but finds no listed equivalency. What does LEED v5 direct?

a) Alternative standards are never permitted in LEED v5
b) Check the project priority library, which lists equivalencies and alternative compliance paths by credit, project type, and region; if none exists, submit a request for USGBC to determine equivalency
c) Petition the standard's publishing organization for a comparability letter
d) Use the alternative standard and document the differences in the submittal narrative

32) Beyond the prerequisite compliance that the standard green lease must require of tenant fit-outs, the Green Leases credit awards achievement for incorporating additional best practices into the lease. Which of the following is one of those listed best practices?

a) Require tenants to purchase renewable energy certificates covering their fit-out construction
b) Ensure brokers or leasing agents have energy training
c) Provide each tenant with a dedicated on-site recycling coordinator
d) Require tenants to complete an annual post-occupancy satisfaction survey

33) Under the Reduce Peak Thermal Loads credit, a LEED BD+C: New Construction project pursues the ventilation energy recovery option. Each fan system supplying outdoor air is fitted with an energy recovery device. To satisfy this option, what minimum recovery performance must each qualifying fan system's recovery system provide?

a) A minimum 75% enthalpy recovery ratio or a minimum 70% sensible heat recovery ratio
b) A minimum 60% enthalpy recovery ratio or a minimum 70% sensible heat recovery ratio
c) A minimum 50% enthalpy recovery ratio or a minimum 75% sensible heat recovery ratio
d) A minimum 70% enthalpy recovery ratio or a minimum 75% sensible heat recovery ratio

34) A new construction team is building its project schedule and wants to engage the commissioning provider at the latest point that still satisfies the Fundamental Commissioning prerequisite. Although early engagement is encouraged so the provider can help shape the owner's project requirements, what is the latest project milestone by which the owner must designate the commissioning provider?

a) By the end of the schematic design phase
b) By the end of the construction documents phase
c) Before predesign begins
d) By the end of the design development phase

35) A LEED project team registered their project in June 2026. In April 2026, USGBC published a LEED interpretation that changes how a prerequisite the team is pursuing should be documented. The team was unaware of this interpretation and has been following the original reference guide approach. What is the team's obligation?

a) The team has no obligation because they were unaware of the interpretation at the time of registration
b) The team should follow the reference guide because it was published before the interpretation
c) The team must comply with the interpretation because it was published before their project registration date
d) The team should request a waiver from GBCI since they were already following the reference guide

36) A design team pursuing the refrigerant path of the Fundamental Refrigerant Management prerequisite specifies a chiller that will be charged with a refrigerant having a 100-year global warming potential of 1,430. According to the prerequisite, what action does this GWP value trigger during the design process?

a) The team must switch to reclaimed refrigerant instead of newly manufactured virgin refrigerant

b) The team is barred from using the equipment because GWP exceeds 1,400

c) The team must develop a list of alternative equipment options with refrigerant GWP less than or equal to 700 and review them with the owner

d) The team must develop alternatives only if the GWP also exceeds 2,088

37) For the design charette required under Integrative Design Process, the charette must be conducted with the owner or owner's representative and participants representing at least how many key perspectives?

a) Three key perspectives

b) Five key perspectives

c) Six key perspectives

d) Four key perspectives

38) A project pursuing the limit-refrigerant-leakage option of Enhanced Refrigerant Management is finalizing its design and installation requirements. Which requirement is stated by this option?

a) Specify an automatic leak detection system in any fully enclosed space housing equipment with an overall refrigerant charge exceeding 100 tCO2e.

b) Specify an automatic leak detection system for every refrigerant-containing unit regardless of charge.

c) Limit all field-installed refrigerant piping to soldered fittings using lead-free solder.

d) Require self-contained equipment only for the units representing at least 50% of the total refrigerant global warming potential.

39) Why is a hotel lobby classified differently from an office lobby in the IEQ space-type definitions?

a) Office lobbies are regularly occupied because of reception staff

b) Hotel lobbies are irregularly occupied like all lobbies

c) The two are classified identically

d) Hotel lobbies are considered regularly occupied because people congregate, work on laptops, and spend more time there than in an office building lobby, which is irregularly occupied

40) Under the Equitable Development credit, a project pursues the equitable construction option using the life-skills training approach rather than apprenticeship training. What scheduling frequency of courses must the project demonstrate?

a) One course per week for the duration of construction

b) One course per month for the duration of construction

c) One course per quarter until the certificate of occupancy

d) One course for every fifty workers hired on the project

41) An office project pursues the plug load management case of the plug and process loads path under Enhanced Energy Efficiency. Which pair of measures does the case require?

a) A plug load dashboard accessible through an application to all regular occupants (with tenant opt-out for display), and IT policies controlling PCs, monitors, and visual displays off when not in use

b) Occupancy-sensor receptacle controls in all private offices, and ENERGY STAR equipment procurement

c) Submetering of all receptacle circuits at 15-minute intervals, and quarterly occupant reporting

d) Smart power strips at all workstations, and after-hours automatic receptacle shutoff

42) Under the Occupant Experience credit, while running the daylight simulation, a team finds several regularly occupied spaces with high annual sunlight exposure. Above what annual sunlight exposure value must the team identify how the space is designed to address glare?

a) ASEnet greater than 20%

b) ASEnet greater than 10%

c) ASEnet greater than 30%

d) ASEnet greater than 25%

43) Under the Fundamental Air Quality prerequisite, a project must provide outdoor airflow measurement devices for its mechanical ventilation systems. Above what outdoor air intake flow does the prerequisite require these airflow measurement devices?

a) Greater than 500 cfm (236 L/s)

b) Greater than 1,000 cfm (472 L/s)

c) Greater than 1,500 cfm (708 L/s)

d) Greater than 2,000 cfm (944 L/s)

44) A team pursuing the daylight simulation path of Occupant Experience asks what minimum average sDA opens recognition. What is the entry threshold?

a) An average sDA of at least 20%

b) An average sDA of at least 75%

c) An average sDA of at least 55%

d) An average sDA of at least 40% across the regularly occupied floor area

45) To satisfy the urban outdoor space requirement of the accessible outdoor space credit, a team designs a pedestrian-oriented plaza as a social area. The building has a daily average occupancy of 400 people including visitors. To qualify as a social area, the space must include seating sufficient to accommodate at least what number of people?

a) 10 people, based on 2.5% of occupants

b) 40 people, based on 10% of occupants

c) 60 people, based on 15% of occupants

d) 20 people, based on 5% of occupants

46) Under the Enhanced Commissioning credit, which standard governs enhanced commissioning of mechanical, electrical, plumbing, control, data center, process, building monitoring, and renewable energy systems?

a) ASHRAE Guideline 0, The Commissioning Process
b) ANSI/ASHRAE/IES Standard 202-2024, Commissioning Process
c) ASTM E2947-21a, Standard Guide for Building Enclosure Commissioning
d) ASHRAE Standard 90.1, Energy Standard for Buildings

47) Under the Fundamental Air Quality prerequisite, which standard governs air filter ratings?

a) ASHRAE 52.2-2017
b) ASHRAE 62.1-2022
c) ASHRAE 241-2023
d) ASHRAE 55-2020

48) A new building will have 42% of its gross floor area without finished interiors at the time of certification. Which rating system must the project use?

a) LEED BD+C: New Construction, because more than half the building is complete
b) LEED O+M: Existing Buildings, because the incomplete spaces will be finished later
c) LEED ID+C: Commercial Interiors, because the incomplete spaces need interior fit-out
d) LEED BD+C: Core and Shell, because at least 40% of the floor area is incomplete

49) Under the Enhanced Refrigerant Management credit, Option 2, how must equipment using refrigerants above 700 GWP be configured?

a) Self-contained, with no field-installed piping
b) Charged with reclaimed refrigerant
c) Housed in a room with leak detection
d) Limited to half the cooling load

50) Under the Building Product Selection and Procurement credit, a product earns a leadership-position score in a single criteria area, representing optimization that peers aspire to achieve. Which achievement level is this, and what multiplier does the product receive?

a) Level 1, earning a 1× multiplier
b) Level 2, earning a 2× multiplier
c) Level 2, earning a 3× multiplier
d) Level 3, earning a 3× multiplier

51) When defining the lighting boundary for the light pollution reduction credit, a public walkway abuts the project's property line. By how far may the property owner move the lighting boundary beyond the property line in this situation?

a) 3 feet (0.9 meters) beyond the property line
b) 10 feet (3 meters) beyond the property line
c) 5 feet (1.5 meters) beyond the property line
d) To the center line of the walkway

52) A project team is preparing temporary no-smoking signage for the construction phase under the construction management prerequisite. Which statement about the no-smoking requirement during construction is false?

a) The temporary signage may be removed once the building shell is weather-tight, before construction completion

b) The temporary signage must be displayed until construction completion

c) The smoking prohibition applies to conventional cigarettes, cannabis, and electronic smoking devices

d) Smoking is prohibited on the entire jobsite except in designated outdoor smoking areas

53) During design review a team debates the difference between the two reflectance metrics used in the Heat Island Reduction credit. Which statement correctly distinguishes solar reflectance index (SRI) from solar reflectance (SR) as the credit describes them?

a) Solar reflectance index measures hardscape heat rejection, whereas solar reflectance measures a roofing material's ability to reject solar heat

b) Solar reflectance index measures a roofing material's ability to reject solar heat, whereas solar reflectance measures the solar heat rejection of hardscape materials

c) Both metrics measure only roofing performance, but solar reflectance index applies to aged materials and solar reflectance applies to new materials

d) Solar reflectance index measures visible light reflected by glass, whereas solar reflectance measures heat absorbed by vegetated surfaces

54) Beyond satisfying the foundational requirements, the Accessibility and Inclusion credit requires a project to implement a minimum number of additional strategies that go beyond the locally applicable accessibility code. What is that minimum count?

a) at least 5 strategies

b) at least 8 strategies

c) at least 10 strategies

d) at least 12 strategies

55) Under the Biodiverse Habitat credit, as part of vegetation restoration, a team must dedicate a pollinator habitat planted with native flowering plants appropriate for local pollinators. What is the minimum total pollinator habitat area required, and what is the minimum size of each planting grouping?

a) At least 100 square feet (9 square meters) total, with plants in groupings of at least 10 square feet (one square meter)

b) At least 110 square feet (10 square meters) total, with plants in groupings of at least 10 square feet (one square meter)

c) At least 110 square feet (10 square meters) total, with plants in groupings of at least 20 square feet (two square meters)

d) At least 215 square feet (20 square meters) total, with plants in groupings of at least 10 square feet (one square meter)

56) A New Construction team pursuing Enhanced Commissioning for the building enclosure is selecting field-testing methods. Which pairing of test and an acceptable referenced standard is correct?

a) Air leakage testing per ASTM E1105.

b) Infrared imaging per ASTM E779.

c) Water penetration testing per ASTM C1153.

d) Water penetration testing per ASTM E1105 or AAMA 501.2.

57) A project team member must distinguish, within the Project Priorities credit, between the project-type pathway and the exemplary performance pathway. Which statement correctly captures the distinction the credit draws between these two pathways?

a) Project-type credits exceed an existing credit's threshold, while exemplary performance addresses geographically specific priorities

b) Project-type credits address unique needs for a given adaptation or building application, while exemplary performance exceeds an existing LEED v5 credit's requirements or next percentage threshold

c) Both pathways require submitting the intent, requirements, submittals, and design approach for a proposed strategy

d) Project-type credits explore future rating-system aspects, while exemplary performance addresses regional importance

58) A project pursues the leak detection sensors option of the Water Metering and Leak Detection credit. The project has an irrigation system, multiple flush fixtures, and two makeup water systems. Which statement correctly describes a requirement of this option?

a) Sensors must cover at least 80% of the project's flush fixtures

b) A single sensor on a makeup line serving both makeup systems satisfies the makeup water requirement

c) The irrigation sensor must be installed at each individual sprinkler head rather than the point of entry

d) Leak detection sensors or meters must be installed on at least 50% of the project's flush fixtures

59) A LEED BD+C: New Construction project pursuing Platinum certification is mapping out how many points it must earn from the various credits. Under the Enhanced Energy Efficiency credit, which is the total point range available and the number of points specifically required for a New Construction Platinum project?

a) 1 to 10 points available, with 8 points required for Platinum

b) 1 to 7 points available, with 5 points required for Platinum

c) 1 to 10 points available, with 10 points required for Platinum

d) 1 to 7 points available, with 8 points required for Platinum

60) A flooring subcontractor asks which products fall within the flooring category of Low-Emitting Materials. Which item is included in the flooring product category?

a) Poured concrete floor slabs
b) Entryway walk-off systems
c) Composite wood subflooring
d) Wet-applied floor coatings

61) A 0.3-acre (0.12-hectare) infill project in the United States is finalizing its erosion and sedimentation control plan. The civil engineer argues that because the disturbed area is well under one acre, the project may follow only the municipality's grading ordinance and skip the U.S. EPA's 2022 Construction General Permit. Under the Minimized Site Disturbance prerequisite, which statement correctly describes the project's obligation?

a) The project must apply the Construction General Permit regardless of size, unless a local equivalent that is equally or more stringent is used
b) The project is exempt from the Construction General Permit because the disturbed area is below one acre (0.4 hectares)
c) The project may follow the local grading ordinance without comparison, because municipal codes automatically satisfy the requirement
d) The project must apply the Construction General Permit only after first obtaining a National Pollutant Discharge Elimination System permit for sites under one acre

62) A project team is reviewing the LEED v5 impact areas to align their project goals. Which of the following correctly lists the three LEED v5 impact areas and their approximate point weightings?

a) Decarbonization (40%), Quality of Life (30%), Ecological Conservation and Restoration (30%)
b) Decarbonization (50%), Quality of Life (25%), Ecological Conservation and Restoration (25%)
c) Decarbonization (50%), Quality of Life (30%), Ecological Conservation and Restoration (20%)
d) Decarbonization (40%), Quality of Life (40%), Ecological Conservation and Restoration (20%)

63) Under Option 2, Fixtures and Fittings, of the Enhanced Water Efficiency credit, how is the baseline for the calculation established?

a) From the maximum flush and flow rates in the prescriptive path of the Minimum Water Efficiency prerequisite
b) From the calculated baseline used in the performance path of the Minimum Water Efficiency prerequisite, documented through the approved calculator
c) From the actual metered consumption of a comparable existing building
d) From the project's Theoretical Irrigation Requirement, scaled to indoor use

64) A Core and Shell project team wants to pursue the thermal environment requirements of the occupant experience credit under ASHRAE Standard 55-2023. Which statement is true about this option's availability?

a) Core and Shell projects may pursue the thermal environment option for 1 point, the same as New Construction.

b) Core and Shell projects do not have a thermal environment option; that topic may instead be addressed in the tenant guidelines.

c) Core and Shell projects may pursue thermal environment for 2 points because tenants are not yet present.

d) Core and Shell projects may pursue thermal environment only if they also pursue the sound environment option.

65) For a project's solar readiness design under the Renewable Energy credit, the main electrical service panel must meet a minimum capacity requirement and include a specially marked reservation. Which statement correctly reflects both requirements?

a) Minimum busbar rating of 100 amps and reserved space marked "For future solar electric"

b) Minimum busbar rating of 200 amps and a pre-installed double pole circuit breaker

c) Minimum busbar rating of 400 amps and reserved space marked "For future solar electric"

d) Minimum busbar rating of 200 amps and reserved space marked "For future solar electric"

66) A Core and Shell project team is preparing its tenant guidelines for Tenant Guidelines. To satisfy the prerequisite, at what point must all tenants receive the tenant guidelines document?

a) After the tenant completes its build-out

b) Before signing the lease

c) At the certificate of occupancy

d) Within one year of move-in

67) Beyond the impact on the completed building's function, the climate resilience assessment must evaluate one additional impact scope for each priority hazard. What is it?

a) The potential impact on the project site during construction

b) The potential impact on adjacent property values

c) The potential impact on the local electric grid

d) The potential impact on national supply chains

68) An ice machine is being specified for a project documenting Minimum Water Efficiency. Which pair of requirements applies?

a) A maximum of 1.3 gpm (4.9 lpm) supply flow and daily flush cycles

b) ENERGY STAR or performance equivalent

c) WaterSense labeling and a water softener

d) Remote refrigeration and potable once-through cooling

69) Under the Renewable Energy credit, a project pursues Tier 1 by installing a renewable system on the site of a social impact project. Which conditions must be met?

a) The social impact entity receives the system at no cost, takes ownership, and keeps the rights to the power; where residents pay their own bills, they must see proportionate savings

b) The LEED project retains ownership of the system and sells the power back to the social impact entity

c) The social impact entity leases the system from the LEED project for a nominal fee

d) The system's renewable energy certificates are retained by the social impact entity for resale

70) A team is weighing whether to prioritize reusing a targeted material type versus an other material type under the materials reuse option of the Building and Materials Reuse credit. What value advantage do targeted materials receive compared to other reused materials?

a) A 1.5× multiplier

b) A 2× multiplier

c) A 3× multiplier

d) A 200% diversion credit

71) A team completes the hazard assessment for the Climate Resilience Assessment prerequisite, identifies its two priority hazards, and files the documentation. A reviewer notes that the prerequisite expects more than a standalone study. Which statement best describes that expectation?

a) The team must implement every resilience strategy identified in the assessment, without exception

b) The assessment satisfies the prerequisite once the hazards are listed and rated; no further application is expected

c) Where possible, the team should use the assessment to inform the planning, design, and operations and maintenance of the project, and describe how project-specific strategies were considered

d) The team must engage a third-party resilience consultant to certify the findings before submittal

72) A team selecting physical diversity strategies under Accessibility and Inclusion asks what the door operator strategy requires. What does the strategy specify?

a) Automatic sliding doors at the main entrance only

b) Revolving doors with an adjacent swing door

c) Wave-to-open or vertical hand or foot press door operators at all regularly used building entrances

d) Push-button operators at accessible entrances only

73) Under Option 2, Monitoring-Based Commissioning, of the Enhanced Commissioning credit, how do the two paths differ?

a) Path 1 uses basic software, and Path 2 uses enhanced software

b) Path 1 covers the envelope, and Path 2 covers MEP systems

c) Path 1 is for New Construction, and Path 2 is for Core and Shell

d) Path 1 requires a one-year commitment, and Path 2 requires five years

74) A healthcare project pursues the submeters option of the Water Metering and Leak Detection credit and has many qualifying water subsystems within its scope. In addition to the standard submetering requirements, in how many of the listed specialized healthcare subsystems must water meters be installed?

a) Any three of the listed subsystems

b) Any five of the listed subsystems

c) Any seven of the listed subsystems

d) All ten of the listed subsystems

75) For data centers and warehouses pursuing the transportation resources path under the Compact and Connected Development credit, one qualifying resource is proximity to a main logistics hub. Within what driving distance of a main logistics hub must the site be located to count that resource?

a) A one-mile (1,600-meter) driving distance

b) A 10-mile (16-kilometer) driving distance

c) A 5-mile (8-kilometer) driving distance

d) A 25-mile (40-kilometer) driving distance

76) A team mapping acoustic expectations under the sound environment option of Occupant Experience is classifying each space's acoustic privacy. Which classification set belongs to acoustic privacy?

a) Loud zone, quiet zone, mixed zone, circulation, and sensitive

b) High speech security, confidential speech privacy, normal speech privacy, marginal speech privacy, or no privacy

c) High risk, medium risk, low risk, or no risk

d) Preserve, improve, restore, mitigate, or specialize

77) Under the Occupant Experience credit, Option 1, Path 2, interior atria may satisfy up to what share of the required view area?

a) 30%

b) 15%

c) 50%

d) 75%

78) Under the Reduce Embodied Carbon credit, Option 1, what must the whole-building life-cycle assessment exclude?

a) Operating energy and water
b) The building enclosure
c) The hardscape
d) Module C impacts

79) Under the Reduce Embodied Carbon credit, Option 2, what must both paths use to demonstrate the reductions?

a) Product-specific or facility-specific Type III EPDs
b) Industry-wide EPDs only
c) Manufacturer marketing data
d) Default baseline values

80) A developer proposes a modular building designed to be relocated to different sites every five years as market conditions change. Is this project eligible for LEED certification?

a) Yes, because modular buildings are encouraged under LEED's material efficiency strategies
b) Yes, but only under LEED O+M: Existing Buildings
c) No, because every LEED project must be in a permanent location on existing land
d) No, because modular buildings cannot meet minimum program requirements for floor area

81) Under the Air Quality Testing and Monitoring credit, a team pursues the continuous indoor air monitoring option. Which set lists exactly the parameters the indoor air monitors are required to measure?

a) Carbon monoxide, PM2.5, ozone, temperature, and relative humidity
b) Carbon dioxide, PM2.5, PM10, formaldehyde, and relative humidity
c) Carbon dioxide, PM2.5, total volatile organic compounds, temperature, and relative humidity
d) Carbon dioxide, total volatile organic compounds, ozone, temperature, and carbon monoxide

82) At a minimum, how much total points are required for a LEED Gold certification?

a) 50 points
b) 59 points
c) 60 points
d) 79 points

83) A wildfire-season heat wave prompts a resilience discussion. What does Resilient Spaces encourage teams to consider about hazard timing?

a) Model only the single most likely event
b) Consider and design for the possibility of multiple events occurring at the same time, such as a wildfire and an extreme heat event
c) Assume hazards alternate seasons
d) Design for one hazard at a time, as simultaneous events are excluded

84) A coastal project addresses tsunamis under the Enhanced Resilient Site Design credit. Which references does the credit direct the team to?

a) ASCE 24 and FEMA 543, because tsunami inundation is treated as flooding

b) The U.S. National Tsunami Hazard Mitigation Program and the TsunamiReady guidelines

c) ASCE/SEI 7 storm-surge provisions and the FORTIFIED Commercial Wind standards

d) The National Wildfire Coordinating Group Standards for Mitigation, applied to evacuation routing

85) After the LEED certification preliminary review, how many appeals can a project file?

a) 5

b) 10

c) 20

d) Unlimited

86) Under the Planning for Zero Waste Operations prerequisite, which material stream is not among those the recycling program must cover?

a) Construction debris

b) Mixed paper

c) Glass

d) Metals

87) A LEED BD+C: Core and Shell project pursues Renewable Energy, Option 2 (Renewable Energy Readiness) by designating a solar zone. Which of the following is not a requirement for the solar readiness solar zone?

a) The solar zone must generate at least 5% of annual base building site energy

b) The dedicated solar zone area must equal at least 40% of the gross roof area

c) The main electrical service panel must have a minimum busbar rating of 200 amps

d) Total solar zone area must be comprised of areas with no dimension less than five feet and no less than 160 square feet

88) A project team registers a new commercial building and wants to know what work it must perform to satisfy the Carbon Assessment prerequisite. Which statement best describes how the 25-year assessment is produced?

a) The project team must hire a third-party consultant to perform a whole-building life-cycle analysis covering a 25-year horizon.

b) The team must independently model annual emissions for each source and submit a comprehensive carbon-accounting workbook to USGBC.

c) USGBC develops the projection, but only after the team enters supplemental refrigerant and transportation data not required elsewhere in the rating system.

d) USGBC develops the 25-year projection using data the team already submitted under the related prerequisites, and the team need not compile any new calculations for this prerequisite.

89) A project sends mixed (commingled) C&D materials to a recovery facility whose recycling rate has not been verified by a USGBC-approved third party. Under Construction and Demolition Waste Diversion, what diversion rate may the project claim for those materials?

a) The facility's self-reported rate of up to 75%
b) A maximum of 35% diversion rate
c) 100%, because the materials left the site for recovery
d) A maximum of 50% diversion rate

90) A New Construction team has met Path 1 and wants to reach the second point of the low-emitting materials credit specifically by adding the furniture category. Under which path does this fall, and what threshold applies to furniture?

a) Path 2, with furniture required at greater than 90% of the furniture product category
b) Path 3, with furniture required at greater than 80% of the furniture product category
c) Path 3, with furniture required at greater than 90% of the furniture product category
d) Path 2, with furniture required at greater than 80% of the furniture product category

91) A Core and Shell project completing Construction and Demolition Waste Diversion asks what must flow to its future tenants. What does the credit require?

a) Nothing; the credit ends at base building completion
b) Require tenants to reuse the base building's waste hauler
c) Provide each tenant a completed waste manifest from the base building work
d) Include the building's approved construction and demolition waste management plan in the tenant guidelines

92) A candidate sorting the strategies of Accessibility and Inclusion asks where nonslip flooring belongs. Which category holds it?

a) Accessibility for navigation
b) Accessibility for physical diversity
c) Accessibility for social health
d) Accessibility for safety and aging

93) Under the Grid Interactive credit, Option 3, Path 1, System-Level Controls, what is required and for what kind of building is it best suited?

a) Automated demand response controls for at least two eligible systems
b) A building automation system plan to shed 10% of peak demand — best suited to large buildings
c) On-site storage sized to the building's peak demand — best suited to any building
d) Enrollment in a demand response program — best suited to campus applications

94) A New Construction project owner is reluctant to commit to the data-sharing provision of the Water Metering and Reporting prerequisite. The LEED AP explains the exact obligation. For how long must the project commit to share whole-project water usage data with USGBC, and under what conditions does the commitment end early?

a) Annually for at least three years, or until a major renovation occurs
b) Monthly for at least five years, regardless of any change in ownership
c) Annually for at least ten years, or until the project is recertified
d) Annually for at least five years, or until the building changes ownership or lessee

95) A team pursuing the previously undeveloped sites option of Sensitive Land Protection is evaluating a parcel where the proposed development footprint would sit 75 feet (23 meters) from a wetland and 75 feet (23 meters) from a pond. Setting aside minor improvements, how should a LEED AP assess the two conditions?

a) Both conditions comply, because both buffers are 50 feet (15 meters)
b) The wetland condition complies, but the pond condition does not, because water bodies carry a 100-foot (30-meter) buffer
c) The pond condition complies, but the wetland condition does not, because wetlands carry a 100-foot (30-meter) buffer
d) Neither condition complies, because both buffers are 100 feet (30 meters)

96) A data center sited near a residential neighborhood pursues the Equitable Development credit and must meet the sensitive project location requirement. What is the minimum distance the project building must maintain from the property lines of the nearest sensitive receptors?

a) 300 feet (90 meters)
b) 200 feet (60 meters)
c) 400 feet (120 meters)
d) 500 feet (150 meters)

97) A team plans to claim the additional point for collecting and reusing a portion of the chosen percentile event volume. Which of the following is not listed as an eligible end use for that reused rainwater under the Rainwater Management credit?

a) Irrigation
b) Flush fixtures
c) Potable drinking water at lavatory faucets
d) Makeup water for cooling towers or boilers

98) For either option of the Enhanced Energy Efficiency credit, which version of ASHRAE Standard 90.1 must the project use?

a) The most recently published version of ASHRAE 90.1
b) The same version of ASHRAE 90.1 referenced in the Minimum Energy Efficiency prerequisite
c) ASHRAE 90.1-2022 in all cases, regardless of registration date
d) Whichever version yields the greater percentage improvement

99) A product manager wants to document Level 1 achievement in the circular economy criteria area under the Building Product Selection and Procurement credit. Which product characteristic qualifies a product for Level 1 in the circular economy criteria area?

a) Disclosure of raw material sources during manufacturing
b) Incorporation of recycled or qualified bio-based content
c) Screening to avoid a wide range of harmful substances
d) Closed-loop circular systems that maximize value through reuse

100) A team conducting a whole-building life-cycle assessment reports that the global warming potential result, normalized to floor area, is 560.8 kg CO2eq per square meter for the product stage. Which life-cycle stage does this normalized result represent for the Quantify and Assess Embodied Carbon prerequisite's reporting?

a) Stages A1 through A5, the product and construction stages combined
b) Stages A through C, the full cradle-to-grave scope
c) Stages B1 through B7, the use stage
d) Stages A1 through A3, the product stage

PRACTICE TEST 2 – ANSWERS

A score above 80 would indicate very well preparation for the exam.

1	C	**26**	C	**51**	C	**76**	B
2	A	**27**	A	**52**	A	**77**	A
3	C	**28**	D	**53**	B	**78**	A
4	D	**29**	A	**54**	C	**79**	A
5	D	**30**	C	**55**	B	**80**	C
6	A	**31**	B	**56**	D	**81**	C
7	B	**32**	B	**57**	B	**82**	C
8	D	**33**	D	**58**	D	**83**	B
9	B	**34**	D	**59**	A	**84**	B
10	B	**35**	C	**60**	B	**85**	D
11	A	**36**	C	**61**	A	**86**	A
12	B	**37**	D	**62**	B	**87**	A
13	B	**38**	A	**63**	B	**88**	D
14	A	**39**	D	**64**	B	**89**	B
15	D	**40**	B	**65**	D	**90**	B
16	C	**41**	A	**66**	B	**91**	D
17	D	**42**	A	**67**	A	**92**	D
18	A	**43**	B	**68**	B	**93**	A
19	A	**44**	D	**69**	A	**94**	D
20	D	**45**	D	**70**	B	**95**	B
21	B	**46**	B	**71**	C	**96**	A
22	B	**47**	A	**72**	C	**97**	C
23	A	**48**	D	**73**	A	**98**	B
24	B	**49**	A	**74**	B	**99**	B
25	B	**50**	B	**75**	B	**100**	D

PRACTICE TEST 2 ANSWERS & EXPLANATIONS

1) Correct answer: C. It is precedent-setting and applies to all LEED projects.

LEED interpretations are official, precedent-setting answers that apply to all LEED projects — not just the one that submitted the inquiry. They are published in the Addenda database and effectively update the guidance for the affected credit. They do not expire. They are mandatory for projects registered after the interpretation is published. This broad applicability distinguishes LEED interpretations from project-specific guidance.

2) Correct answer: A. The product must have been tested no more than three years prior to the product's purchase.

For a qualifying independent laboratory report, the product must have been tested no more than three years prior to the product's purchase, and the report must declare testing according to CDPH Standard Method v1.2-2017 with the test date being less than three years from the date of purchase.

3) Correct answer: C. An inspection within 24 hours of the occurrence of a storm event that produces 0.25 inches (6 millimeters) or more of rain within a 24-hour period.

Minimized Site Disturbance requires site inspections of all controls and management practices at least once every seven calendar days, or once every 14 calendar days and within 24 hours of the occurrence of a storm event that produces 0.25 inches (6 millimeters) or more of rain within a 24-hour period. A team choosing the 14-day interval is therefore obligated to also inspect within 24 hours after a qualifying storm. Separately, dewatering inspections must occur once per day on which the discharge of dewatering occurs.

4) Correct answer: D. Reduce, Reuse, Recycle, Compost, Anaerobic Digestion.

Planning for Zero Waste Operations instructs teams to follow the true waste hierarchy, which ranks strategies from most to least preferred in this order: Reduce (minimize waste at the source), Reuse (extend product life), Recycle (transform waste into new products), Compost (decompose organics into nutrient-rich soil), and Anaerobic Digestion (break down organic material into a nutrient-rich liquid). The framework prioritizes reducing generation and maximizing resource recovery before disposal.

5) Correct answer: D. Building-Level Water Metering.

Option 1 lists the prerequisites tenant-installed systems must comply with as the Climate Resilience Assessment, Human Impact Assessment, Carbon Assessment, Minimum Water Efficiency, Minimum Energy Efficiency, Fundamental Commissioning, Energy Metering and Reporting, Planning for Zero Waste Operations, and Fundamental Air Quality.

6) Correct answer: A. Floor decking, roof decking, and enclosure.

The maintained structure includes floor decking, roof decking, and the enclosure. The other groupings are partial or unrelated.

7) Correct answer: B. A higher threshold of 1.8 COP and a lower threshold of 1.3 COP.

For Core and Shell projects on the no on-site combustion path, the combined weighted average equipment efficiency for space heating and service water heating must be at least 1.8 coefficient of performance for 4 points and at least 1.3 coefficient of performance for 3 points.

8) Correct answer: D. one storage space per three dwelling units.

For residential projects, long-term storage must serve at least 15% of all regular building occupants but no fewer than one storage space per three dwelling units, in addition to the short-term spaces. The four-spaces-per-building floor is the commercial, institutional, schools, and healthcare minimum, not the residential one, so it does not apply here.

9) Correct answer: B. 75%.

The Operable Windows option scales with coverage: designing operable windows for 50% of the regularly occupied spaces earns 1 point, while 75% earns 2 points. The windows must also satisfy the opening size and location requirements of the natural ventilation procedure in ASHRAE 62.1-2022, Section 6.4.

10) Correct answer: B. 30 spaces.

The readiness option requires residential projects to make EV-ready the greater of 20% of total parking or 20 spaces. For 150 spaces, 150 x 0.20 = 30 spaces, which exceeds the 20-space floor, so 30 EV-ready spaces are required. The 15-space answer applies a 10% figure, which is the commercial readiness percentage, not the residential one.

11) Correct answer: A. At least 75%.

For Path 2: Connected Site, an infill site requires that at least 75% of the project boundary border parcels that are previously developed sites, earning 2 points. By contrast, an adjacent site requires that at least a contiguous 25% of the project boundary border previously developed parcels, earning 1 point. Bordering rights-of-way do not themselves constitute previously developed land; instead, it is the status of the property on the other side of the right-of-way that counts. Any part of the boundary that borders a water body is excluded from the calculation.

12) Correct answer: B. Emissions from electricity decline toward zero over time, while emissions from on-site combustion remain constant.

The projection assumes the grid emissions factor declines across the analysis period as generation decarbonizes, so emissions attributable to electricity fall toward zero. Emissions from on-site combustion, by contrast, are produced at the building by burning fuel and stay constant — no external trend reduces them. This asymmetry is the entire argument of the prerequisite: a combustion-based building locks in emissions that the grid's improvement will never erase, which is precisely why the requirement pushes teams toward electrification or a written plan to retrofit away from fuel.

13) Correct answer: B. Only projects registered before January 1, 2028, may use Option 1; Option 2 is required for projects registered on or after that date.

For Minimum Energy Efficiency, projects registering before January 1, 2028, may comply with either Option 1 (ASHRAE 90.1-2019 with Addendum cr) or Option 2 (ASHRAE 90.1-2022). Projects registering on or after January 1, 2028, must comply with Option 2.

14) Correct answer: A. By exceeding the credit requirements or achieving the next incremental percentage threshold.

Exemplary performance earns points by exceeding the credit requirements or achieving the next incremental percentage threshold for the credit, for credits identified as eligible in USGBC's Project Priority Library.

15) Correct answer: D. Five years (or until the building changes ownership or lessee); monthly energy consumption for each source, on-site renewable energy generation, and peak electrical demand.

Under Energy Metering and Reporting, the project must commit to reporting to USGBC at least annually the monthly energy data for 12 consecutive months of total energy consumption for each energy source, on-site renewable energy generation, and peak electrical demand. This commitment must carry forward for five years or until the building changes ownership or lessee. Note this reporting period (five years) is distinct from the requirement that the monitoring system be capable of maintaining all collected data for 36 months.

16) Correct answer: C. 1.28 gallons per flush (4.8 liters per flush).

Under the prescriptive path, the maximum installed flush rate for a toilet (water closet) is 1.28 gallons per flush (4.8 liters per flush). The value of 1.6 gallons per flush (6.0 liters per flush) is the baseline figure used in the performance-path calculation, not a permissible installed rate, so it would fail the prescriptive table.

17) Correct answer: D. Smaller, less complex projects are not required to have an energy model.

To make LEED v5 more accessible, smaller, less complex projects are not required to have an energy model for the Design Analysis. Teams may instead analyze efficiency, peak load reduction, and decarbonization measures using simplified energy modeling, analysis from similar projects, or analysis from published data.

18) Correct answer: A. 300 feet (90 meters); no fewer than four spaces.

Long-term bicycle storage must be provided within 300 feet (90 meters) walking distance from any functional entry, but no fewer than four storage spaces per building, in addition to the short-term spaces.

19) Correct answer: A. A whole-building air leakage pressurization test, or a continuous air barrier design and installation verification program.

The standard requires an independent third party to test whole-building pressurization for air leakage; alternately, a continuous air barrier design and installation verification program conducted by an independent third party satisfies the requirement.

20) Correct answer: D. They have high impacts such as embodied carbon or pollution, are hard to recycle, and significant amounts of them end up in landfill.

Under Building and Materials Reuse, the materials reuse option lists specific targeted materials that are valued higher in the calculations because they have high impacts — embodied carbon or pollution — are hard to recycle, and significant amounts of them end up in landfill. Reuse is their best exit from the waste stream. The targeted materials are carpeting, ceilings, furniture (ancillary and systems), and interior walls.

21) Correct answer: B. Only once, even though it is addressed through multiple strategies.

The credit states plainly that each surface may only be counted once, even if it is addressed through multiple strategies. This prevents teams from inflating the weighted sum by layering a shade structure over high-reflectance paving and claiming both.

22) Correct answer: B. SMACNA IAQ Guidelines for Occupied Buildings under Construction, 2nd edition, 2007, ANSI/SMACNA 008-2008, Chapter 3.

The construction management practices are primarily adapted from the SMACNA IAQ Guidelines for Occupied Buildings under Construction, 2nd edition, 2007, ANSI/SMACNA 008-2008, Chapter 3. ASHRAE 62.1-2022 and ASHRAE 52.2-2017 are referenced by the Fundamental Air Quality prerequisite for ventilation and filter testing, not by this prerequisite, and the SMACNA HVAC Duct Construction Standards address ductwork fabrication rather than construction-phase IAQ.

23) Correct answer: A. Prohibit all smoking on-site, with signage posted on the property line.

School projects must prohibit all smoking on-site—not merely within a setback—and signage must be posted on the property line to indicate the no-smoking policy.

24) Correct answer: B. The Mindful Materials Common Materials Framework and the AIA Architecture and Design Materials Pledge.

Under Building Product Selection and Procurement, the five criteria areas — climate health, human health, ecosystem health, social health and equity, and circular economy — align with the impact areas of the Mindful Materials Common Materials Framework (CMF) and the AIA Architecture and Design (A&D) Materials Pledge. Both initiatives standardize product evaluation and function independently of specific certifications, providing a structure that lets various ecolabels and standards map to consistent impact areas, scored by disclosure, verification, and optimization.

25) Correct answer: B. Those stages must be included in the analysis as appropriate, even though A1 through A3 are zeroed.

For reused or salvaged materials in a whole-building life-cycle assessment, teams zero out the manufacturing life-cycle stages A1 through A3, but the other impacts—transport (A4) and the B through C stages—must still be included in the analysis as appropriate. Reuse eliminates the upfront extraction and manufacturing burden, not the emissions of moving the material or its eventual end-of-life handling. Treating the material as entirely carbon-free, doubling the transport stage, or excluding later stages at will all contradict the rule. A further constraint is that baseline and proposed quantities must remain the same for reused materials.

26) Correct answer: C. The project must process at least 120,000 lb (57,606 kg) of laundry per year.

To use the commercial washing machine table, the project must process at least 120,000 lb (57,606 kg) of laundry per year. The 2,400 lb (10,886 kg) value is real but refers to the minimum machine capacity per 8-hour shift for the compliant on-premise washer specification, not an annual processing threshold or an annual capacity. The 100-meals trigger belongs to the commercial kitchen equipment table, not laundry, so the loads-per-day option conflates two different tables.

27) Correct answer: A. 0.25 miles (400 meters).

To credit proximity to an emergency cooling station as an extreme heat resilience element, the station must be located within 0.25 miles (400 meters) of the building for easy access. This walkable distance is what makes the refuge usable by vulnerable populations during a heat event.

28) Correct answer: D. 1.15.

The increased ventilation option requires a 15% increase in breathing-zone outdoor air ventilation rates above the minimum determined under the fundamental air quality prerequisite for the single point. For a single-zone or 100% outdoor air system, the calculated minimum outdoor airflow for the system is multiplied by 1.15.

29) Correct answer: A. On-site electricity generation and fuel combustion cannot be used to meet the demand-side management criteria.

Under Grid Interactive, Option 3: Automated Demand-Side Management, the rating system states that on-site electricity generation and fuel combustion cannot be used to meet the demand-side management criteria—this expressly includes renewable electricity generation, which is separately credited in Renewable Energy.

30) Correct answer: C. The lighting boundary may move to the center line of that street.

When a property line abuts a public street, alley, or transit corridor, the lighting boundary may move to the center line of that street, alley, or corridor — recognizing that the project is responsible for only its half of the public way. The 5-foot (1.5-meter) offset is the neighboring rule for public areas such as a walkway, bikeway, plaza, or parking lot that abuts the property line, and confusing the two is the most common error. The lighting boundary is typically based on property lines but expressly may extend beyond them under these defined conditions, so it does not always coincide with the LEED project boundary.

31) Correct answer: B. Check the project priority library, which lists equivalencies and alternative compliance paths by credit, project type, and region; if none exists, submit a request for USGBC to determine equivalency.

LEED v5 is written on the most current, widely available standards, but alternatives may be acceptable. The project priority library includes a list of equivalencies and alternative compliance paths available by credit, project type, or region. Where no guidance allows an equivalency, teams submit a request for USGBC to determine equivalency.

32) Correct answer: B. Ensure brokers or leasing agents have energy training.

Ensuring brokers or leasing agents have energy training is one of the listed best practices that can be written into the green lease. Others include requiring the tenant to pay for their own electric and nonelectric energy and water use, implementing a cost recovery clause for energy-efficiency upgrades benefiting the tenant, disclosing tracked common area energy and water use, disclosing the whole-building ENERGY STAR or locally applicable equivalent score, requiring energy or water efficiency fit-outs that improve upon the minimum efficiency prerequisites, metering tenant use beyond what the metering prerequisites require, limiting on-site combustion emissions, and implementing indoor air quality or thermal comfort best practices.

33) Correct answer: D. A minimum 70% enthalpy recovery ratio or a minimum 75% sensible heat recovery ratio.

The ventilation energy recovery option requires each fan system supplying outdoor air to have an energy or heat recovery system with a minimum 70% enthalpy recovery ratio or a minimum of 75% sensible heat recovery ratio. Provisions must also exist to bypass or control the recovery system during moderate outdoor air conditions, and in aggregate, fan systems supplying less than 15% of the project's total outdoor air may be excluded.

34) Correct answer: D. By the end of the design development phase.

Under the Fundamental Commissioning prerequisite, the owner must designate the commissioning provider by the end of the design development phase; the provider is encouraged to engage earlier (in predesign) to assist with the owner's project requirements, but design development is the latest acceptable milestone.

35) Correct answer: C. The team must comply with the interpretation because it was published before their project registration date.

Projects are required to adhere to all LEED interpretations published before their registration date, regardless of whether the team was aware of them at the time. Since this interpretation was published in April 2026 and the project registered in June 2026, the interpretation applies to the project and compliance is mandatory.

36) Correct answer: C. The team must develop a list of alternative equipment options with refrigerant GWP less than or equal to 700 and review them with the owner.

Under the refrigerant option of this prerequisite, whenever specifications reference equipment with refrigerant GWP exceeding 700 (or do not specify the refrigerant), the team must develop a list of alternative equipment options with refrigerant GWP less than or equal to 700 and review all proposed options with the owner. A GWP of 1,430 exceeds 700, so the evaluation is triggered.

37) Correct answer: D. Four key perspectives.

Integrative Design Process requires that, during predesign or early in design, the team conduct a charette with the owner or owner's representative and participants representing at least four key perspectives (for example, architect, contractor, energy modeler, and community engagement representatives).

38) Correct answer: A. Specify an automatic leak detection system in any fully enclosed space housing equipment with an overall refrigerant charge exceeding 100 tCO2e.

The limit-refrigerant-leakage option requires an automatic leak detection system in fully enclosed spaces housing equipment with an overall refrigerant charge exceeding 100 tCO2e. Detection is not mandated for every unit regardless of charge; field-installed piping must use brazed or press-type fittings (not lead-free solder); and self-contained equipment is required for at least 80% (not 50%) of the total refrigerant global warming potential.

39) Correct answer: D. Hotel lobbies are considered regularly occupied because people congregate, work on laptops, and spend more time there than in an office building lobby, which is irregularly occupied.

In the EQ category's space-type guidance, lobbies are generally irregularly occupied — but hotel lobbies are regularly occupied, because people often congregate, work on laptops, and spend more time there than in an office building lobby.

40) Correct answer: B. One course per month for the duration of construction.

The life-skills training approach under equitable construction requires programs conducted on the construction site covering topics such as financial literacy or first-time home buying, demonstrating scheduling of one course per month for the duration of construction.

41) Correct answer: A. A plug load dashboard accessible through an application to all regular occupants (with tenant opt-out for display), and IT policies controlling PCs, monitors, and visual displays off when not in use.

Under Enhanced Energy Efficiency, the plug load management case requires exactly two measures. First, a plug load dashboard accessible through an application to all regular building occupants. Tenants may opt out of displaying their loads to other tenants. Second, for building or tenant types with IT departments, policies for PCs, monitors, and visual displays to be controlled off when not in use, except during scheduled maintenance periods.

42) Correct answer: A. ASEnet greater than 20%.

For any regularly occupied space whose net annual sunlight exposure (ASEnet, evaluated at 1000 lux for 250 hours) is greater than 20%, the team must identify how the space is designed to address glare. The credit states that based on IES experience an ASEnet of 20% or higher is a level of concern, prompting architectural mitigation or an automated daylight management system. Importantly, even when ASE exceeds this level, those spaces are not excluded from the average sDA calculation that determines points.

43) Correct answer: B. Greater than 1,000 cfm (472 L/s).

The prerequisite requires outdoor airflow measurement devices for all mechanical ventilation systems with an outdoor air intake flow greater than 1,000 cfm (472 L/s); monitoring is encouraged for smaller systems but not required.

44) Correct answer: D. An average sDA of at least 40% across the regularly occupied floor area.

Under Occupant Experience, the daylight simulation path's recognition begins at an average sDA of 40% for the total regularly occupied floor area, climbing through 55%, 65%, and 75%. The research bands give the scale meaning: 55% to 75% is nominally acceptable daylight, and 75% or higher is preferred. Averages are calculated without excluding any space for high annual sunlight exposure — spaces above 20% ASE instead document their glare response.

45) Correct answer: D. 20 people, based on 5% of occupants.

A social area must include sufficient seating to accommodate at least 5% of the daily average occupants, including visitors. Five percent of 400 occupants is 20 people. A social area is one of four ways to satisfy the urban outdoor space requirement, alongside a biophilic space, a garden, or a recreational area, and the space must be accessible to users of all abilities and located near building entrances or amenities.

46) Correct answer: B. ANSI/ASHRAE/IES Standard 202-2024, Commissioning Process.

The credit references ANSI/ASHRAE/IES Standard 202-2024, Commissioning Process, for enhanced commissioning of mechanical, electrical, plumbing, control, data center, process, building monitoring, and renewable energy systems. The strongest distractor is ASTM E2947-21a, which the credit also references — but for building enclosure commissioning, not MEP. ASHRAE 90.1 is the energy standard rather than a commissioning standard.

47) Correct answer: A. ASHRAE 52.2-2017.

Air filter performance (the MERV rating method) is governed by ASHRAE 52.2-2017. ASHRAE 62.1 is the ventilation standard, 241 covers air cleaning, and 55 covers thermal comfort.

48) Correct answer: D. LEED BD+C: Core and Shell, because at least 40% of the floor area is incomplete.

LEED BD+C: Core and Shell is required when at least 40% of the project's gross floor area is incomplete at certification — meaning it lacks basic finishes or essential MEP systems. With 42% incomplete, this project exceeds the 40% threshold and cannot use New Construction (which requires at least 60% complete). O+M applies to buildings operational for at least one year. ID+C applies to interior renovations with no exterior scope.

49) Correct answer: A. Self-contained, with no field-installed piping.

Field-installed piping is the most common leak path, so high-GWP equipment (above 700 GWP) must be self-contained with no field-installed piping. The other options are plausible refrigerant measures but not what Option 2 requires.

50) Correct answer: B. Level 2, earning a 2× multiplier.

The credit defines three achievement levels: Level 1 (a first step toward sustainability, earning a 1× multiplier), Level 2 (a leadership position in the marketplace that peers aspire to achieve, earning a 2× multiplier), and Level 3 (elite, forefront-of-sustainability products, earning a 3× multiplier).

51) Correct answer: C. 5 feet (1.5 meters) beyond the property line.

When a public area such as a walkway, bikeway, plaza, or parking lot abuts the property line, the property owner may move the lighting boundary to 5 feet (1.5 meters) beyond the property line.

52) Correct answer: A. The temporary signage may be removed once the building shell is weather-tight, before construction completion.

The prerequisite requires no-smoking signage to be displayed until construction completion, so removing it once the shell is weather-tight is the false statement. The other choices are true: the prohibition covers conventional cigarettes (cigarettes, cigars, pipes), cannabis (medical or recreational), and electronic smoking devices, and smoking is banned on the entire jobsite during construction except in designated outdoor smoking areas.

53) Correct answer: B. Solar reflectance index measures a roofing material's ability to reject solar heat, whereas solar reflectance measures the solar heat rejection of hardscape materials.

For nonroof measures, SR values are used; for roofs, SRI values are used.

54) Correct answer: C. at least 10 strategies.

The credit requires projects to include at least 10 of the accessibility and inclusion strategies most relevant to the project that go beyond the locally applicable accessibility code, drawn from the four tables covering physical diversity, safety and aging, social health, and navigation. If fewer than 10 are relevant because of project-type variations, teams may submit up to three alternative inclusive design strategies that meet the intent of the credit.

55) Correct answer: B. At least 110 square feet (10 square meters) total, with plants in groupings of at least 10 square feet (one square meter).

The credit requires at least 110 square feet (10 square meters) consisting of native flowering plants appropriate for local pollinators, with plants placed in groupings of at least 10 square feet (one square meter), and the pollinator area must be designated with signage. The 100-square-foot value is a rounded near-miss that does not match the stated minimum, the 20-square-foot grouping doubles the actual grouping floor, and 215 square feet (20 square meters) doubles the total area. Signage must include information and education about the habitat's purpose and the native species planted.

56) Correct answer: D. Water penetration testing per ASTM E1105 or AAMA 501.2.

Enhanced commissioning for the building enclosure requires water penetration testing per ASTM E1105 or AAMA 501.2. Air leakage testing uses ASTM E783, E779, E1186, or E3158, and infrared imaging uses ASTM C1153 or ASTM C1060, so the other pairings assign standards to the wrong test.

57) Correct answer: B. Project-type credits address unique needs for a given adaptation or building application, while exemplary performance exceeds an existing LEED v5 credit's requirements or next percentage threshold.

Project-type credits are identified as addressing unique needs for the given adaptation or building application, whereas exemplary performance earns points by exceeding the credit requirements or achieving the next incremental percentage threshold of an existing LEED v5 credit.

58) Correct answer: D. Leak detection sensors or meters must be installed on at least 50% of the project's flush fixtures.

The leak detection sensors option requires installing flow meters or sensors on at least 50% of the project flush fixtures identified in the minimum water efficiency prerequisite; a device may serve an individual toilet room or a group restroom. The 80% figure is the submetering threshold for indoor fixtures under the other option. A single sensor on a shared makeup line serving multiple systems does not meet the credit, which requires leak detection at each makeup water system. Irrigation devices must be installed at the point of entry to the site, not at each sprinkler head.

59) Correct answer: A. 1 to 10 points available, with 8 points required for Platinum.

For LEED BD+C: New Construction, the Enhanced Energy Efficiency credit offers a range of 1 to 10 points, of which 8 points are required for New Construction Platinum projects.

60) Correct answer: B. Entryway walk-off systems.

Under Low-Emitting Materials, the flooring category covers nonstructural flooring by surface area or cost. It includes hard and soft finishes, raised flooring systems, entryway (walk-off) systems, area rugs, wood subflooring, underlayments, and sandwich panels. It also includes air and vapor barrier membranes used inside an air barrier membrane.

Poured concrete is excluded from the low-emitting product categories altogether. Composite wood subflooring migrates to the composite wood category. Wet-applied floor products belong to the paints and coatings or adhesives and sealants categories.

61) Correct answer: A. The project must apply the Construction General Permit regardless of size, unless a local equivalent that is equally or more stringent is used.

The plan must conform to the erosion and sedimentation requirements of the U.S. EPA's 2022 Construction General Permit (CGP), the EU Taxonomy pollution-prevention provisions, or a local equivalent, and the prerequisite is explicit that projects must apply the CGP regardless of size.

62) Correct answer: B. Decarbonization (50%), Quality of Life (25%), Ecological Conservation and Restoration (25%).

LEED v5 is organized around three impact areas: Decarbonization accounts for approximately 50% of available points, Quality of Life accounts for approximately 25%, and Ecological Conservation and Restoration accounts for approximately 25%.

63) Correct answer: B. From the calculated baseline used in the performance path of the Minimum Water Efficiency prerequisite, documented through the approved calculator.

This option further reduces fixture and fitting water use from the calculated baseline established in the performance path of the Minimum Water Efficiency prerequisite, and compliance for both the prerequisite and this credit is documented through the USGBC-approved calculator. The baseline flush and flow rates come from the prerequisite's baseline table while the design case reflects the fixtures actually installed, with the total number of uses held constant between the two. Metered consumption from another building plays no role, and the Theoretical Irrigation Requirement is the outdoor baseline rather than an indoor one.

64) Correct answer: B. Core and Shell projects do not have a thermal environment option; that topic may instead be addressed in the tenant guidelines.

Core and Shell projects do not have options for the thermal or sound environment, because there are fewer opportunities to address these topics within the Core and Shell scope; the credit directs that these topics may instead be included in the tenant guidelines.

65) Correct answer: D. Minimum busbar rating of 200 amps and reserved space marked "For future solar electric".

The solar readiness electrical service criteria require the main electrical service panel to have a minimum busbar rating of 200 amps and a reserved space permanently marked "For future solar electric" to allow installation of a future double pole circuit breaker.

66) Correct answer: B. Before signing the lease.

Tenant Guidelines (a Core and Shell-only prerequisite) requires the base building to create tenant guidelines to be shared with all tenants before signing the lease. Distributing them after build-out, at the certificate of occupancy, or within a year of move-in would defeat the intent, because the guidelines are meant to ensure tenants consider sustainability during their design and build-out phases.

67) Correct answer: A. The potential impact on the project site during construction.

Under Climate Resilience Assessment, each priority hazard is assessed for its potential impact on the project site and building function and its potential impact on the project site during construction.

68) Correct answer: B. ENERGY STAR or performance equivalent.

Under Minimum Water Efficiency, ice machines must be ENERGY STAR or performance equivalent and must use either air-cooled or closed-loop cooling, such as a chilled or condenser water system.

69) Correct answer: A. The social impact entity receives the system at no cost, takes ownership, and keeps the rights to the power; where residents pay their own bills, they must see proportionate savings.

For a social impact project, the renewable power system must be provided, installed, and commissioned at no cost to the social impact entity, ownership must be transferred to that entity, and the rights to the power must be given to it. Where residents pay their own bills, they must see proportionate savings.

70) Correct answer: B. A 2× multiplier.

The credit states that targeted materials receive a 2× multiplier compared to other reused materials because they are valued higher for their high impacts, difficulty to recycle, and tendency to end up in landfill. The targeted materials are carpeting, ceilings, furniture (ancillary and systems), and interior walls.

71) Correct answer: C. Where possible, the team should use the assessment to inform the planning, design, and operations and maintenance of the project, and describe how project-specific strategies were considered.

The prerequisite states that, where possible, teams should use the information from the assessment to inform the planning, design, and operations and maintenance of the project and describe how project-specific strategies were considered. The two priority hazards are identified specifically to be addressed through proposed design strategies, so the assessment is meant to shape decisions rather than sit as a standalone report. It is not, however, a mandate to implement every identified strategy — the requirement is to consider and describe them.

72) Correct answer: C. Wave-to-open or vertical hand or foot press door operators at all regularly used building entrances.

Under Accessibility and Inclusion, the physical diversity category includes providing wave-to-open or vertical hand or foot press door operators at all regularly used building entrances. Touchless and limb-flexible actuation serves users who cannot grip, twist, or push conventional hardware — and the coverage is all regularly used entrances, not a single accessible door. Strategies must exceed the applicable local code to count toward the credit's required ten.

73) Correct answer: A. Path 1 uses basic software, and Path 2 uses enhanced software.

Under Option 2, projects can implement monitoring-based commissioning with basic software under Path 1 or with enhanced software under Path 2. The distinction is the sophistication of the monitoring technology, with Path 2 layering advanced analytics on top of the Path 1 foundation.

The paths are not defined by building system, rating system, or commitment length — the multi-year monitoring commitment applies to both.

74) Correct answer: B. Any five of the listed subsystems.

For healthcare projects under the submeters option, the credit requires installing water meters in any five of the listed subsystems, such as purified water systems, filter backwash water, dietary, laundry, laboratory, central sterile, hydrotherapy, surgical suite, closed-loop hydronic makeup, or cold-water makeup for domestic hot water. If a project does not include five of the additional subsystems within its scope, it may instead submeter all applicable subsystems.

75) Correct answer: B. A 10-mile (16-kilometer) driving distance.

Under the transportation resources path, a site qualifies if it is within a 10-mile (16-kilometer) driving distance of a main logistics hub. Other resources, such as a highway on-off ramp or a freight rail access point, use a one-mile (1,600-meter) driving distance.

76) Correct answer: B. High speech security, confidential speech privacy, normal speech privacy, marginal speech privacy, or no privacy.

Under Occupant Experience, the mapping path classifies each primary indoor and outdoor space across several categories. Acoustic privacy uses high speech security, confidential speech privacy, normal speech privacy, marginal speech privacy, or no privacy. The other sets are real classifications for the neighboring categories. Loud, quiet, mixed, circulation, and sensitive zones classify acoustic comfort. High through no risk classifies noise exposure zones. Preserve, improve, restore, mitigate, and specialize classify soundscape management.

77) Correct answer: A. 30%.

Interior atria may be used to meet up to 30% of the required view area, so most of the views must still be to the true exterior environment.

78) Correct answer: A. Operating energy and water.

The assessment covers the structure, enclosure, and hardscape but excludes operating energy and water, since those are addressed by the energy and water credits.

79) Correct answer: A. Product-specific or facility-specific Type III EPDs.

Both Option 2 paths require product-specific or facility-specific Type III environmental product declarations to substantiate the claimed reductions.

80) Correct answer: C. No, because every LEED project must be in a permanent location on existing land.

One of LEED's minimum program requirements states that every project must be in a permanent location on existing land. A project designed to move at any point in its lifetime is not eligible for LEED certification. While modular construction techniques are valued for material efficiency, the building itself must remain permanently on its site. This requirement is not related to floor area minimums or specific rating systems — it is a fundamental eligibility criterion for all LEED projects.

81) Correct answer: C. Carbon dioxide, PM2.5, total volatile organic compounds, temperature, and relative humidity.

The continuous monitoring option requires monitors for exactly five parameters: carbon dioxide, PM2.5, total volatile organic compounds, temperature, and relative humidity.

82) Correct answer: C. 60 points.

According to the total number of points a project receives, LEED certification levels are determined:

LEED Certified: 40–49 points (This is the base certification level.)

LEED Silver: 50–59 points

LEED Gold: 60–79 points

LEED Platinum: 80+ points

83) Correct answer: B. Consider and design for the possibility of multiple events occurring at the same time, such as a wildfire and an extreme heat event.

Under Resilient Spaces, the credit addresses episodic disruptive events — extreme weather, wildfires, pandemics, power outages — through separate options. The reference guide encourages teams to consider and design for the possibility of multiple events occurring at the same time, such as a wildfire and an extreme heat event. Smoke argues for closing the building while heat argues for ventilation, so compound events stress designs single-hazard thinking would pass.

84) Correct answer: B. The U.S. National Tsunami Hazard Mitigation Program and the TsunamiReady guidelines.

For tsunamis, the credit references the U.S. National Tsunami Hazard Mitigation Program, or a local equivalent, and additionally requires teams to incorporate elements from the TsunamiReady guidelines, such as installing danger-area and evacuation signage. Tsunami is treated as its own hazard with its own references, so applying the flood standards ASCE 24 and FEMA 543 is incorrect even though both involve water. ASCE/SEI 7 and the FORTIFIED standards address wind, and the wildfire mitigation standards are unrelated.

85) Correct answer: D. Unlimited.

Projects can file unlimited number of appeals during the LEED certification process.

86) Correct answer: A. Construction debris.

The recycling program must cover organics/food waste, mixed paper, corrugated cardboard, glass, plastics, and metals. Construction debris is an operations-phase omission — it belongs to the Construction and Demolition Waste Diversion credit, not this prerequisite.

87) Correct answer: A. The solar zone must generate at least 5% of annual base building site energy.

Renewable Energy, Option 2: Renewable Energy Readiness (Core and Shell only) is a design readiness option and does not require any actual energy generation, so a 5% of annual base building site energy generation threshold is not a solar zone requirement; that figure is the New Construction Tier 1 one-point threshold of 5% of annual site energy under Option 1, whereas the Core and Shell Tier 1 one-point threshold is 15% of annual base building site energy. The genuine solar zone requirements include a dedicated solar zone of at least 40% of the gross roof area, a main electrical service panel with a minimum busbar rating of 200 amps, and total area comprised of areas with no dimension less than five feet and no less than 160 square feet.

88) Correct answer: D. USGBC develops the 25-year projection using data the team already submitted under the related prerequisites, and the team need not compile any new calculations for this prerequisite.

USGBC will provide the project team with a 25-year projection of the project's emissions, drawing on data the team already submitted under the Operational Carbon Projection and Decarbonization Plan prerequisite, the Fundamental Refrigerant Management prerequisite, and the Quantify and Assess Embodied Carbon prerequisite, and it may also include the Transportation Demand Management credit if the project pursues it.

89) Correct answer: B. A maximum of 35% diversion rate.

Under Construction and Demolition Waste Diversion, mixed C&D materials sent to a processing facility must use the facility's average recycling rate, but recycling rates not verified by a third party must assume a maximum of 35% diversion rate. The 35% cap reflects an approximate average U.S. facility recycling rate and applies whenever a USGBC-approved third party (such as the Recycling Certification Institute) has not certified the facility. Source-separated materials, by contrast, are considered 100% diverted, and salvaged off-site materials are valued at twice the diversion rate (200%).

90) Correct answer: B. Path 3, with furniture required at greater than 80% of the furniture product category.

Earning the second point by adding furniture on top of Path 1 is Path 3, which requires the furniture category at greater than 80% of the furniture product category. The credit separates this from Path 2, which instead reaches the second point through any two of adhesives and sealants, walls, insulation, or composite wood.

91) Correct answer: D. Include the building's approved construction and demolition waste management plan in the tenant guidelines.

Under Construction and Demolition Waste Diversion, Core and Shell projects must include the building's approved construction and demolition waste management plan in the tenant guidelines — the pre-lease document created under the Tenant Guidelines prerequisite. Tenant fit-outs generate their own construction waste, and handing tenants the working plan extends the diversion infrastructure into their scope.

92) Correct answer: D. Accessibility for safety and aging.

Under Accessibility and Inclusion, nonslip flooring belongs to the safety and aging category, alongside fixed area rugs with transition strips, glazing indication, audible and visual emergency alerts, closed risers, visual contrast, and photoluminescent floor-level warnings. Physical diversity covers movement and access strategies; social health covers restrooms, lactation, changing stations, and language; navigation covers wayfinding. Sorting strategies to categories is how the credit's ten-strategy count is assembled.

93) Correct answer: A. Automated demand response controls for at least two eligible systems.

Path 1, System-Level Controls, requires automated demand response controls for at least two eligible systems — from HVAC, lighting, automatic receptacle controls, service water heating, and electric vehicle supply equipment — and is best suited to smaller buildings without a building automation system. Controlling individual systems directly is the practical approach when there is no central automation to orchestrate load shedding. The building-automation-system plan to shed 10% is Path 2, and storage and program enrollment are other options entirely.

94) Correct answer: D. Annually for at least five years, or until the building changes ownership or lessee.

Under Water Metering and Reporting, the project owner must commit to sharing the resulting whole-project water usage data with USGBC at least annually, and this commitment must carry forward for five years or until the building changes ownership or lessee. Data may be shared using a USGBC-approved template or an approved third-party source such as ENERGY STAR Portfolio Manager.

95) Correct answer: B. The wetland condition complies, but the pond condition does not, because water bodies carry a 100-foot (30-meter) buffer.

Under Sensitive Land Protection, the development footprint may not sit within 100 feet (30 meters) of a water body or within 50 feet (15 meters) of a wetland, except for minor improvements. At 75 feet, the footprint clears the wetland buffer but violates the pond buffer. The water body buffer is exactly double the wetland buffer.

96) Correct answer: A. 300 feet (90 meters).

Under the sensitive project location requirement for data centers, warehouses, and distribution centers, the project building must be located a minimum of 300 feet (90 meters) from the property lines of the nearest sensitive receptors, such as residential areas, schools, day care centers, and hospitals.

97) Correct answer: C. Potable drinking water at lavatory faucets.

The credit lists the eligible end uses for reused rainwater as irrigation, flush fixtures, makeup water systems (e.g., cooling towers or boilers), and other process water demands.

98) Correct answer: B. The same version of ASHRAE 90.1 referenced in the Minimum Energy Efficiency prerequisite.

For both options, the project must use the same version of ASHRAE 90.1 referenced in the Minimum Energy Efficiency prerequisite. Keeping the credit aligned with the prerequisite ensures the additional efficiency is measured against the same baseline and edition the project already established for compliance — a project cannot claim savings against one edition of the standard while documenting the prerequisite against another.

99) Correct answer: B. Incorporation of recycled or qualified bio-based content.

Products rewarded under Level 1 of circular economy include those that have incorporated recycled or qualified bio-based content. Preserving materials through reuse and maximizing value through closed-loop circular systems is reserved for Levels 2 and 3 of this same criteria area. Disclosing raw material sources is a Level 1 ecosystem health behavior, and screening to avoid harmful substances belongs to Levels 2 or 3 of human health, so both are near-miss distractors from sibling criteria areas.

100) Correct answer: D. Stages A1 through A3, the product stage.

When reporting for the prerequisite from a whole-building life-cycle assessment, teams include the data for product stages A1 through A3, and the sample results table lists the global warming potential of 560.8 kg CO2eq per square meter under exactly that product stage. The prerequisite is a cradle-to-gate exercise, so it does not extend to construction stages A4–A5, to the full cradle-to-grave A–C scope, or to the B1–B7 use stage. Those broader boundaries belong to the Reduce Embodied Carbon credit's whole-building life-cycle assessment, not to this prerequisite's reporting.

SECTION 4:

PRACTICE TEST 3

1) A zero lot line project pursuing the percentile of rainfall events option of Rainwater Management asks how its retention thresholds compare to standard projects. What are the zero lot line percentile tiers?

a) The 60th, 70th, and 80th percentile events, in place of the standard 80th, 85th, and 90th
b) The 50th, 60th, and 70th percentile events, in place of the standard 75th, 85th, and 95th
c) The same 80th, 85th, and 90th percentile events as all projects
d) The 70th, 75th, and 80th percentile events, in place of the standard 80th, 85th, and 90th

2) A residential unit includes a wood-burning fireplace that is neither closed combustion nor power-vented. What does Fundamental Air Quality require for it?

a) Removal of the fireplace before certification
b) Solid glass enclosures or doors that seal when closed
c) A carbon monoxide monitor within the firebox room only
d) An outside combustion air duct only

3) Under the Renewable Energy credit, what does a single renewable energy certificate represent?

a) The environmental benefits associated with one megawatt-hour of renewable electricity generation

b) One kilowatt of installed renewable generating capacity

c) One year of operation of an on-site renewable system

d) A one-percent reduction in the project's annual site energy use

4) Under the Enhanced Air Quality credit, which standard's measures do both options draw on?

a) ASHRAE Guideline 42-2023

b) ASHRAE 62.1-2022

c) ASHRAE 52.2-2017

d) ASHRAE 241-2023

5) Under the Enhanced Energy Efficiency credit, which of the following best describes nonprocess energy, also called regulated energy?

a) Equipment tied to building use and operations, such as computers, kitchen appliances, and elevators

b) Building systems that condition spaces and provide comfort and basic amenities, such as interior and exterior lighting, HVAC, hot water heating, toilet exhaust, and parking garage ventilation

c) Only the energy consumed by on-site renewable generation systems

d) The upstream energy lost in generating and delivering electricity to the building

6) When quantifying the global warming potential of structure, enclosure, and hardscape materials under Quantify and Assess Embodied Carbon, the prerequisite specifies a minimum set of materials that must be included. Which grouping correctly reflects that minimum list?

a) Only concrete, structural steel, and insulation, because these are the three highest-impact materials

b) Concrete, steel, gypsum board, carpeting, ceilings, and interior partition framing

c) Asphalt, concrete, masonry, structural steel, insulation, aluminum extrusions, structural wood and composites, cladding, and glass

d) Plumbing fixtures, mechanical equipment, elevators, and conveying systems within the project boundary

7) Under the Reduce Parking path of the Transportation Demand Management credit, which parking is excluded when calculating the project's total provided parking capacity against the baseline?

a) Off-street spaces leased by the project that lie outside the project boundary

b) On-street parking located in public rights-of-way

c) New off-street spaces owned by the project within the project boundary

d) Existing off-street spaces used by the project but owned by a third party

8) Under the Enhanced Commissioning credit, what must the commissioning provider verify with respect to operator training in both enhanced commissioning paths?

a) That the training curriculum was approved by the local authority having jurisdiction

b) That operator training met the training plan and actually took place

c) That operators passed a certification exam administered by the manufacturer

d) That training is scheduled to occur within one year after occupancy

9) A facade uses bronze-tinted vision glass, and the team asks whether views through it can qualify under the quality views path of Occupant Experience. What does the path provide?

a) Any tint disqualifies the glazing

b) Tints are acceptable only below 20% VLT

c) Only perfectly clear glass qualifies

d) Views must be through glass with visible light transmittance above 40%, and where glazing has frits, patterns, or tints, the view must be preserved

10) A LEED BD+C: New Construction project team with several long-term commercial tenants wants to pursue the Green Leases credit by writing sustainability obligations into its tenant leases. What is the outcome?

a) The credit may be pursued, but the achievement is capped at a single point for New Construction

b) The credit may be pursued only if more than 40% of the gross floor area will be tenant-occupied

c) The credit is not available, because the Green Leases credit applies only to Core and Shell projects

d) The credit may be pursued only if the team also satisfies the Tenant Guidelines prerequisite

11) A project team installs solar photovoltaic canopies over its surface parking lot. Under the Heat Island Reduction credit, how does this single design decision interact with other credits?

a) It counts only toward the Renewable Energy credit, because energy generation systems are excluded from heat island calculations

b) It counts as a nonroof measure under the Heat Island Reduction credit by providing shade, and separately supports the Renewable Energy credit by generating on-site power

c) It counts only toward the Heat Island Reduction credit, because on-site generation used for shading cannot also be claimed as renewable energy

d) It counts toward neither, because parking canopies are addressed exclusively under the parking under cover option

12) A team pursuing the Heat Island Reduction credit, Option 1, plans to install a vegetated roof. What does the credit require of that installation?

a) That it achieve a minimum aged solar reflectance index equivalent to a high-reflectance roof

b) That it be irrigated by a permanent system

c) That it be installed using native or adapted plant species

d) That it cover the entire applicable roof area to be claimed

13) A team planning deconstruction asks who should be involved in the salvage assessment under Building and Materials Reuse. What does LEED recommend?

a) A licensed auctioneer

b) The waste hauler's dispatcher

c) Only the design architect

d) Salvage experts such as deconstruction contractors who know how to safely remove materials and which materials carry salvage value

14) A multifamily New Construction project is designing dwelling-unit safety provisions under Fundamental Air Quality. What does the prerequisite require for carbon monoxide monitoring?

a) Battery-only CO monitors in common corridors

b) A single CO monitor at each building entrance

c) CO monitors only in units containing combustion appliances

d) A CO monitor on each floor of each dwelling unit, hard-wired with battery backup, in all unit types regardless of the equipment installed

15) Under the Light Pollution Reduction credit, which of the following is not among the exterior lighting that may be exempt from the uplight and light trespass requirements?

a) Lighting for hospital emergency departments, including associated helipads

b) Lighting for theatrical purposes for stage, film, and video performances

c) Decorative lighting for a private rooftop terrace used for evening events

d) Government-mandated roadway lighting

16) Under the Transportation Demand Management credit, to count toward the bicycle network, on-street bicycle facilities must be located on streets with a maximum posted speed limit of no more than:

a) 20 mph (32 km/h)

b) 30 mph (48 km/h)

c) 25 mph (40 km/h)

d) 35 mph (56 km/h)

17) A residential project pursuing the location-efficient affordable housing path provides affordable rental dwelling units under the Equitable Development credit. For how long must those rental units be maintained at affordable levels, and which units are excluded from the requirement calculations?

a) A minimum of 10 years, with existing dwelling units excluded from the calculations

b) A minimum of 15 years, with all units including existing ones counted in the calculations

c) A minimum of 30 years, with existing dwelling units excluded from the calculations

d) A minimum of 15 years, with existing dwelling units excluded from the calculations

18) Under Option 1 of the Green Leases credit, the standard green lease must include tenant requirements ensuring that tenant-installed systems comply with a specified set of prerequisites. Which of the following is among that set?

a) Fundamental Refrigerant Management

b) Quantify and Assess Embodied Carbon

c) No Smoking or Vehicle Idling

d) Planning for Zero Waste Operations

19) Under Option 4, Peak Thermal Load Reductions, of the Reduce Peak Thermal Loads credit, the three paths quantify the peak load reduction differently. Which pairing of path to method is correct?

a) Path 1 uses ASHRAE 90.1 Appendix G to calculate a performance index

b) Path 1 caps peak load intensity using the Passive House tools, WUFI Passive or the Passive House Planning Package

c) Path 2 uses the Passive House Planning Package to cap peak load intensity

d) Path 3 demonstrates percentage improvements through the ASHRAE 90.1-2022 envelope trade-off analysis and the Total System Performance Ratio method

20) A transportation engineer setting the parking baseline for the reduce parking path of Transportation Demand Management asks which reference governs. What does the path specify?

a) The national average of one space per 250 square feet (23 square meters)

b) The project's own traffic study without external reference

c) Base ratios from the Institute of Transportation Engineers Parking Generation Manual, sixth edition, or a comparable resource applied by a qualified transportation engineer or planner, or supplementary LEED guidance

d) The local zoning ordinance's minimum parking requirement

21) A new building will have 48% of its gross floor area without finished interiors at the time of certification. Which rating system must the project use?

a) LEED BD+C: New Construction, because more than half the building is complete

b) LEED O+M: Existing Buildings, because the incomplete spaces will be finished later

c) LEED ID+C: Commercial Interiors, because the incomplete spaces need interior fit-out

d) LEED BD+C: Core and Shell, because at least 40% of the floor area is incomplete

22) While documenting the Fundamental Refrigerant Management prerequisite, a reviewer asks which standard governs the global warming potential values cited for each refrigerant in the inventory. Which basis does the prerequisite specify for these GWP figures?

a) 20-year GWPs assessed per the IPCC Fifth Assessment Report

b) 100-year GWPs taken solely from the EPA AIM Act Technology Transitions Rule

c) 100-year GWPs assessed in accordance with the IPCC Fourth Assessment Report or later

d) 500-year GWPs assessed under the Kigali Amendment to the Montreal Protocol

23) Materials salvaged and sent off-site to reuse markets count toward which credit?

a) Construction and Demolition Waste Diversion

b) Building and Materials Reuse

c) Building Product Selection and Procurement

d) Reduce Embodied Carbon

24) A team pursues a regional priority strategy under the Project Priorities credit. In addition to the general supporting documentation, what must the team specifically include for the regional priority strategy?

a) A description of the regional or local priority, with evidence and description of strategies adopted by the project to address it

b) Evidence of pilot credit registration and a completed survey

c) A summary of the significant, measurable environmental performance improvement proposed

d) Confirmation that a LEED AP with a Building Design and Construction specialty is on the team

25) Under Renewable Energy, what distinguishes Tier 2 (New Off-site Renewable Electricity) from Tier 3 (Off-site Renewable Energy) qualified electricity generators?

a) Tier 2 must be located on the project's contiguous campus, while Tier 3 may be anywhere in the country

b) Tier 2 requires Green-e Energy certification, while Tier 3 does not

c) Tier 2 requires a 10-year contract, while Tier 3 allows a 1-year contract

d) Tier 2 must be operational within two years of occupancy or have a commercial operations date no more than five years before the contract, while Tier 3 electricity may be up to fifteen years old

26) Under Option 1, Path 2, of the Enhanced Energy Efficiency credit, which statement correctly describes Case 3 and how the cases may be combined?

a) Case 3 uses the ASHRAE 90.1 exceptional calculation method to demonstrate a percentage improvement in plug, process, refrigeration, and conveyance loads; Cases 1 and 2 may be combined, but Case 3 is pursued on its own

b) Case 3 requires a plug load dashboard; all three cases must be pursued together

c) Case 3 uses the exceptional calculation method; it may be freely combined with Cases 1 and 2

d) Case 3 addresses only refrigeration loads; it may not be combined with any other case, and neither may Cases 1 and 2

27) A warehouse project asks which Location and Transportation credits offer it project-type-specific content. Which pairing is correct?

a) Equitable Development adds the sensitive project location requirement, and Compact and Connected Development adds the surrounding development and resources option

b) Sensitive Land Protection and Electric Vehicles both add warehouse-specific options

c) No LT credit differentiates warehouses

d) Transportation Demand Management adds a freight-routing path

28) A project team is generating the annual site energy estimate for the Operational Carbon Projection and Decarbonization Plan prerequisite. Which approach is acceptable?

a) Only a calibrated whole-building energy model may be used

b) Teams running an energy model for the Minimum Energy Efficiency prerequisite may take the values from the simulation results; teams relying on similar projects or published data may estimate from predicted design conditions

c) Teams must use metered data from the first year of operation, submitted retroactively

d) Teams must use the default energy use intensity values published for the project's climate zone

29) A general contractor is drafting the extreme heat protection portion of its construction management plan. The prerequisite's extreme heat requirement is adapted from guidance issued by which organization, and which of the following is an example of a preventive measure the plan should address?

a) Adapted from OSHA heat-prevention guidance; measures include providing cool, shaded, or air-conditioned rest areas, required rest breaks, and scheduling labor-intensive work for cooler parts of the day

b) Adapted from the SMACNA IAQ guidelines; measures include sealing doorways and tenting high-activity areas to isolate emissions

c) Adapted from ASHRAE Standard 55; measures are limited to maintaining the indoor comfort zone within the mechanically conditioned space

d) Adapted from the WHO global air quality guidelines; measures include lowering the interior particulate design limit during hot months

30) A team is siting a new construction project on previously undeveloped land that includes a natural lake and a small natural wetland near the property edge. To avoid encroaching on sensitive land under the previously undeveloped sites option, what are the minimum buffer distances the development footprint must respect for the water body and the wetland, respectively (apart from allowed minor improvements)?

a) 100 feet (30 meters) from the water body and 50 feet (15 meters) from the wetland

b) 50 feet (15 meters) from the water body and 50 feet (15 meters) from the wetland

c) 50 feet (15 meters) from the water body and 100 feet (30 meters) from the wetland

d) 100 feet (30 meters) from the water body and 100 feet (30 meters) from the wetland

31) A large office with a building automation system pursues the building automation system path of the automated demand-side management option under the Grid Interactive credit. The team must develop a plan to automatically shed a portion of the project's peak electricity demand. What minimum shed amount and duration must the plan address?

a) At least 5% of peak electricity demand for a minimum of one hour

b) At least 10% of peak electricity demand for a minimum of three hours

c) At least 10% of peak electricity demand for a minimum of one hour

d) At least 20% of peak electricity demand for a minimum of one hour

32) Under Occupant Experience, Option 5, Path 2 (Quality Electric Lighting), which set of criteria correctly states an acceptable way to satisfy the electric light glare control and color rendering requirements for regularly occupied spaces?

a) A unified glare rating (UGR) of 25 or lower, and a color rendering index of at least 80.

b) Calculated luminance of less than 6,000 cd/sq. m. between 45 and 90 degrees from nadir, and a color rendering index of at least 90.

c) Calculated luminance of less than 10,000 cd/sq. m. between 0 and 45 degrees from nadir, and an Rf of 90 or higher.

d) A spatial daylight autonomy of at least 55%, and a gamut index (Rg) of 78 or higher.

33) A team wants the first point available under the diversion component of the construction and demolition waste diversion credit. Which combination of thresholds must the project meet to earn that single point?

a) Divert at least 75% of total material, with at least 25% of diverted materials salvaged or source-separated to single-material recyclers

b) Divert at least 50% of total material, with at least 25% of diverted materials salvaged or source-separated to single-material recyclers

c) Divert at least 50% of total material, with at least 10% of diverted materials salvaged or source-separated to single-material recyclers

d) Divert at least 60% of total material, with at least 10% of diverted materials salvaged or source-separated to single-material recyclers

34) A project is aiming for the maximum two points under the diversion component of the construction and demolition waste diversion credit. The team diverts 78% of total construction and demolition material. Of the diverted materials, what minimum share must be salvaged or source-separated and sent to single-material recyclers to secure both points?

a) 25%

b) 10%

c) 20%

d) 50%

35) A team documenting composite wood under Low-Emitting Materials submits a third-party certification showing the product complies with the CDPH Standard Method v1.2-2017. Does this satisfy the composite wood category?

a) Yes; CDPH testing satisfies every product category

b) No; composite wood uses the formaldehyde emissions evaluation criteria, such as ULEF or NAF certification under EPA TSCA Title VI or CARB ATCM, not the CDPH method

c) No; composite wood requires an environmental product declaration instead

d) Yes, provided the private office scenario was modeled

36) Under the Electric Vehicles credit, a commercial New Construction project with 80 total parking spaces intends to pursue one point through the electric vehicle readiness option only. What is the minimum number of EV-ready spaces it must design?

a) 8 spaces
b) 16 spaces
c) 20 spaces
d) 10 spaces

37) For each priority hazard, the Climate Resilience Assessment prerequisite lists specific items the project team must assess and specify. Which of the following is not one of those required items?

a) The projected service life of the project
b) Potential impact on the project site during construction
c) A cost-benefit analysis comparing the proposed resilience strategies against projected avoided losses
d) Exposure, sensitivity, adaptive capacity, vulnerability, and overall risk levels

38) A team completing the occupant experience category of its Human Impact Assessment asks what content belongs there. Which set of factors does the category cover?

a) Adjacent transportation infrastructure, diverse uses, and applicable accessibility codes
b) Housing affordability, access to social services, and community safety
c) Race and ethnicity, income, employment rate, and household types
d) Opportunity for daylight, views, and operable windows; environmental conditions of air and water; and adjacent soundscapes, lighting, and wind patterns

39) Under the Minimum Energy Efficiency prerequisite, a team uses the Energy Cost Budget method. Against what is the proposed design compared?

a) A fixed baseline building defined by the standard, independent of the proposed design
b) A code-minimum version of the same building
c) The median performance of similar buildings in the same climate zone
d) The project's own prior-year metered consumption

40) A U.S. project plans to remove existing invasive vegetation and replant disturbed areas. The landscape architect wants to confirm the standard governing which new plantings are acceptable. Under the Minimized Site Disturbance prerequisite, planted vegetation must be native or adapted to which reference?

a) The project's EPA Level III ecoregion, or a local equivalent for projects outside the U.S.
b) The project's EPA Level IV ecoregion, or a local equivalent for projects outside the U.S.
c) The USDA plant hardiness zone for the project's location
d) Any region within the same U.S. state as the project

41) A new construction project subject to the ASHRAE Standard 90.1 provisions is specifying a monitoring system that can create user reports for consumption and demand at hourly, daily, monthly, and annual intervals. For how long must this system be capable of maintaining all data it collects to meet the Energy Metering and Reporting prerequisite?

a) 12 months

b) 24 months

c) 36 months

d) 60 months

42) A LEED project achieves 81 points, fully electrifies all building systems, procures 100% renewable energy, and achieves 8 points in enhanced energy efficiency. The project also reduces embodied carbon by 15%. What certification level does this project receive under LEED v5 BD+C: New Construction?

a) Platinum, because it exceeds 80 points

b) Platinum, because it meets three of four decarbonization requirements

c) Gold, because the embodied carbon reduction falls below the required 20% threshold for Platinum

d) Silver, because failing any decarbonization requirement drops the project two levels

43) A multifamily residential building pursues the Water Metering and Leak Detection credit under Option 1: Submeters, and plans to install a permanent water meter for each dwelling unit. Which statement correctly describes the requirement for those residential unit meters?

a) The meters must be utility owned and utility grade to qualify for the credit

b) The meters need not be utility owned or utility grade

c) The meters must be read only by the utility provider and reported to USGBC monthly

d) A single meter for the entire building satisfies the residential requirement

44) While planning the design charette required by the Integrative Design Process credit, a facilitator consults the U.S. Department of Energy's Handbook for Planning and Conducting Charrettes for High-Performance Projects. What role does that document play for this credit?

a) It is the credit's referenced standard, and compliance with it is mandatory

b) It establishes the minimum number of hours the design charette must run

c) It is a supporting resource that supplies practical checklists and agendas for facilitating well-structured meetings, workshops, and charrettes

d) It defines the four key perspectives that must be represented among charette participants

45) Under the Enhanced Refrigerant Management credit, Option 1, Path 2, the total weighted average refrigerant GWP must not exceed what share of the GWP benchmarks for the base point?

a) 50%

b) 70%

c) 80%

d) 90%

46) Within the accessible outdoor space credit, a portion of the required outdoor space must be vegetated. What is the minimum vegetated share of the required outdoor space, and how must that vegetation be characterized?

a) At least 20% of the required outdoor space, planted with two or more types of vegetation or with an overhead vegetative canopy

b) At least 30% of the required outdoor space, planted with a single native species

c) At least 25% of the required outdoor space, planted with two or more types of vegetation or with an overhead vegetative canopy

d) At least 25% of the total site area, planted with two or more types of vegetation

47) Under the Grid Interactive credit, which requirement applies to every option?

a) Projects must install on-site electric storage regardless of the option chosen

b) Projects must evaluate the grid-interactive measures against the current and forecasted grid context, and provide interval recording meters and equipment capable of accepting an external signal

c) Projects must enroll in a demand response program under any option

d) Projects must island critical equipment for at least 72 hours under any option

48) A manufacturer's product demonstrates achievement in exactly two of the criteria areas under the Building Product Selection and Procurement credit. How does the credit classify such a product?

a) As an optimized product

b) As a Level 2 product

c) As a multi-attribute product

d) As a circular product

49) A team is preparing the performance path fixture calculations for the Minimum Water Efficiency prerequisite and wants to modify the default gender assumption used for occupancy. Under what condition may the default ratio be changed?

a) The default gender mix is two-thirds female, and any change must be approved by the local water utility

b) The default gender mix is half male and half female, and modifications must be shown to apply for the life of the building

c) The default gender mix is set by the manufacturer of the installed fixtures and cannot be altered

d) The default gender mix is half male and half female, but modifications are prohibited under any circumstances

50) For the previously undeveloped sites option, the credit lists specific referenced standards teams may consult to verify sensitive land criteria within the United States. Which set of standards does the credit reference?

a) Natural Resources Conservation Service, NatureServe, and the U.S. Endangered Species Act

b) Federal Emergency Management Agency, U.S. Fish and Wildlife Service, and the Clean Water Act

c) U.S. Geological Survey, the Ramsar Convention, and the National Environmental Policy Act

d) Environmental Protection Agency, the International Union for Conservation of Nature, and NatureServe

51) A team pursues the regional priority pathway of the Project Priorities credit for an office building in a coastal, flood-prone city. According to the credit, where must the team source a credit so that it qualifies as a regional priority credit?

a) From the project's local LEED Regional Council credit catalog

b) From the GBCI Credential Maintenance Program list

c) From the Arc platform's regional adaptation registry

d) From USGBC's Project Priority Library

52) A LEED BD+C: New Construction project has three team members who each hold a LEED AP with a Building Design and Construction specialty. How many points can the project earn under the LEED AP credit?

a) 1 point

b) 2 points

c) 3 points

d) 4 points

53) Which credit addresses waste generated during the construction phase, as distinct from the Planning for Zero Waste Operations prerequisite?

a) Construction and Demolition Waste Diversion

b) Building and Materials Reuse

c) Quantify and Assess Embodied Carbon

d) Building Product Selection and Procurement

54) Under the Reduce Embodied Carbon credit, Option 2, Path 1, the whole-project weighted average embodied carbon is compared against:

a) Industry average values

b) The prior year's design

c) A net-zero benchmark

d) The regional building code

55) Which phase of the building's life does the Planning for Zero Waste Operations prerequisite address?

a) The operations phase
b) The construction phase
c) The demolition phase
d) The predesign phase

56) Under the low-emitting criteria of Low-Emitting Materials, how are salvaged or reused materials that are more than one year old treated with respect to the VOC emissions evaluation?

a) They automatically comply with the VOC emissions evaluation and do not require emissions evaluations
b) They must still be retested using CDPH Standard Method v1.2-2017 before installation
c) They are excluded from the credit entirely because they are not new products
d) They may only count if they carry a valid third-party certification at the time of purchase

57) A project pursuing Option 4, Outdoor Water Use, of the Enhanced Water Efficiency credit uses the efficient irrigation path. What reduction must the project demonstrate?

a) A reduction of at least 30% in the Theoretical Irrigation Requirement
b) A reduction of at least 40% in the Theoretical Irrigation Requirement
c) A reduction of at least 50% in the Theoretical Irrigation Requirement
d) A reduction of at least 65% in total outdoor water consumption

58) A 120,000-square-foot (11,150-square-meter) office building conducts preoccupancy air testing under Air Quality Testing and Monitoring. Using the credit's table, how many measurement locations are required?

a) Ten
b) Seven
c) Four
d) Twelve

59) Under the Occupant Experience credit, Option 1, Path 2, view glazing must have a visible light transmittance above:

a) 40%
b) 30%
c) 50%
d) 25%

60) A New Construction team intends to use as-designed material quantities for both its baseline and its final results under the Reduce Embodied Carbon credit. Between design and construction, the quantity of one major material increased by 14%. What does the credit require of the team?

a) The team must update the analysis because the quantities changed more than 10% from design through construction

b) Nothing; as-designed quantities may always be used regardless of how much they change

c) The team must switch entirely to as-constructed quantities only if the change exceeds 25%

d) The team must recalculate only the baseline, leaving the proposed design at as-designed quantities

61) A stair core is being detailed under the safety and aging category of Accessibility and Inclusion. What does the strategy for floor-level changes require?

a) Carpet on all stair treads

b) Painted yellow nosings on every third step

c) Visual, tactile, contrasting, or photoluminescent warnings at floor-level changes

d) Audible chimes at each landing

62) The achievement-level definitions and scoring tables of Building Product Selection and Procurement will change over time. Through what mechanism are they updated?

a) Annual republication of the reference guide

b) The LEED addenda process

c) A ballot of LEED AP credential holders

d) Each project team's discretion

63) Under the Biodiverse Habitat credit, under the bird-friendly glass option, the team must assess glass within defined vertical distances. For glass measured from grade and for glass above a green roof, which pair of height limits applies?

a) Up to 40 feet (12 meters) from grade, and up to 20 feet (six meters) from the finished grade of a green roof

b) Up to 50 feet (15 meters) from grade, and up to 15 feet (five meters) from the finished grade of a green roof

c) Up to 60 feet (18 meters) from grade, and up to 30 feet (nine meters) from the finished grade of a green roof

d) Up to 50 feet (15 meters) from grade, and up to 20 feet (six meters) from the finished grade of a green roof

64) A gym under the physical diversity category of Accessibility and Inclusion incorporates accessible exercise equipment. What else does the fitness strategy require?

a) A dedicated adaptive-fitness staff member

b) Reserved hours for users with disabilities

c) An open and accessible route to and around the equipment

d) Video monitoring of all equipment

65) The commissioning team under the Fundamental Commissioning prerequisite must include verification and testing providers. What does a verification and testing provider do?

a) Develops the owner's project requirements on the owner's behalf

b) Carries out the functional performance testing and confirms that portions of the project meet their stated requirements

c) Approves contractor submittals in place of the design team

d) Issues the certificate of occupancy following successful testing

66) Under the Enhanced Refrigerant Management credit, Option 3 is available only to projects where food retailing is what share of gross area?

a) At least 10%

b) More than 20%

c) At least 40%

d) More than 50%

67) A LEED BD+C: Core and Shell project pursues Option 2, Electrification Readiness, of the Electrification credit. Which set of provisions does this option require?

a) Fully installed electric heating and service water heating equipment achieving a weighted average of at least 1.8 coefficient of performance

b) Dedicated equipment space, chase ways, junction boxes and dedicated electrical panel capacity, and accommodations supporting energy recovery ventilation for at least 50% of the ventilation air, with the details documented in the tenant guidelines

c) On-site renewable generation sized to offset the future electrified heating load

d) A signed commitment from each future tenant to install heat pumps within five years of lease execution

68) Under the Enhanced Commissioning credit, when must the commissioning provider be engaged?

a) During predesign or very early in the design phase

b) No later than the start of construction

c) At schematic design phase

d) During the discovery phase

69) An architect designing circulation paths under the baseline accessibility provisions of Accessibility and Inclusion asks about passing spaces on narrow paths. What do the provisions require?

a) Passing spaces of 48 inches (1.22 meters) at least every 100 feet (30 meters) along any path

b) No passing spaces, provided the path is one-way

c) Passing spaces of 60 inches (1.525 meters) at least every 200 feet (61 meters) along any path less than 60 inches (1.525 meters) wide

d) Passing spaces of 60 inches (1.525 meters) at least every 300 feet (91 meters) along any path less than 72 inches (1.83 meters) wide

70) A team pursuing Option 2, Ventilation Energy Recovery, of the Reduce Peak Thermal Loads credit is applying the requirement to its outdoor air fan systems. Which statement correctly describes the requirement?

a) One central fan system with energy recovery satisfies the option for the whole building

b) Every outdoor air fan system must include energy recovery at a 70% enthalpy or 75% sensible recovery ratio, with bypass controls for moderate conditions; systems supplying less than 15% of the total outdoor air may be excluded in aggregate

c) Every outdoor air fan system must include energy recovery at a 50% sensible recovery ratio, with no exclusions

d) Energy recovery is required only on systems serving densely occupied spaces above a defined occupancy threshold

71) Under the Reduce Embodied Carbon credit, Option 2, Path 1, from where do the industry average values come?

a) U.S. EPA or the Carbon Leadership Forum Material Baselines

b) ASHRAE 90.1 appendices

c) The manufacturer's warranty data

d) The project's own prior EPDs

72) The tenant guidelines must designate a point of contact. Who may serve in this role, and what is that person responsible for?

a) A representative from the base building team or the building's owner, responsible for coordinating communication about the base building's sustainable design and construction documentation.

b) An independent third-party commissioning agent, responsible for verifying tenant fit-out compliance.

c) A USGBC-assigned reviewer, responsible for approving each tenant's build-out before construction.

d) The lead tenant's facility manager, responsible for distributing guidelines to the remaining tenants.

73) Under the Minimum Water Efficiency prerequisite, a project in Canada is purchasing residential clothes washers, a residential dishwasher, and an ice machine. Which labeling program applies to this equipment?

a) WaterSense, since all water-using products fall under that label

b) ENERGY STAR

c) The Consortium for Energy Efficiency tiers

d) No label applies to appliances; only fixtures and fittings are regulated

74) Under the light trespass requirements of Light Pollution Reduction, what determines the maximum allowed backlight rating for a given luminaire?

a) The MLO lighting zone together with the luminaire's mounting location

b) The lamp's correlated color temperature

c) The MLO lighting zone together with the luminaire's mounting location and its distance from the lighting boundary, measured in mounting heights

d) The MLO lighting zone together the luminaire's height above grade in feet

75) A green building project team is applying for a LEED certification under the LEED BD+C: New Construction and Major Renovations rating system and has chosen the "split review" option. In which of the following phases, the project team cannot appeal the results?

a) After GBCI issues the final certification report

b) After the final design review by GBCI

c) During the preliminary review by GBCI

d) After the final construction review by GBCI

76) A project's climate resilience assessment identified flooding and hurricane winds as its highest priority hazards, but the team wants to pursue the respiratory disease management mode and operable windows options of Resilient Spaces. Is this permitted?

a) Only for Core and Shell projects, which are exempt from hazard alignment

b) No; Resilient Spaces options must address the two highest priority hazards identified in the assessment

c) Only if the team amends its climate resilience assessment to add respiratory disease

d) Yes; teams may select any two options or paths, even those not identified as high priorities in the climate resilience assessment, though assessment findings should inform selection

77) A project pursues the submeters option of the Water Metering and Leak Detection credit. For the installed submeters, what is the minimum data-recording capability and the required mode of data access for the facility manager and tenants?

a) Capable of recording at least daily, with access via monthly reports

b) Capable of recording at least every 15 minutes, with access in real time via building management system only

c) Capable of recording at least hourly, with access in real time via local network, building management system, cloud service, app, or online database

d) Capable of recording at least monthly, with access via monthly utility bills

78) LEED charges two types of fees during the certification process: a registration fee and a review fee. How does the review fee differ from the registration fee?

a) The review fee is a flat fee while the registration fee varies by project size

b) The review fee is based on the project's gross floor area and membership status, while the registration fee is a flat fee

c) Both fees are identical flat fees regardless of project characteristics

d) The review fee is paid to USGBC while the registration fee is paid to Prometric

79) A project owner has agreed to share whole-project water usage data with USGBC to satisfy the Water Metering and Reporting prerequisite. The team must confirm an acceptable method for transmitting the data. Which method of data sharing is expressly named as acceptable for this commitment?

a) A signed annual affidavit from the facility manager summarizing consumption for the calendar year

b) A USGBC-approved data template or an approved third-party data source such as ENERGY STAR Portfolio Manager

c) The utility provider's monthly billing statements forwarded directly to the local green building council chapter

d) A publicly posted dashboard hosted on the project owner's corporate website updated in real time

80) Under the Occupant Experience credit, Option 4, Path 1, across how many categories must acoustic expectations be mapped?

a) Five

b) Three

c) Four

d) Six

81) An office tower pursuing the quality views path of Occupant Experience wants to count workstations facing a planted interior atrium. To what extent may atrium views count toward the required area?

a) Atrium views count fully, the same as exterior views

b) Views into interior atria may meet up to 50% of the required area

c) Atrium views never count toward the requirement

d) Views into interior atria may meet up to 30% of the required area

82) Every project in any LEED rating system will be scored out of 100 points. However, there are also 10 additional bonus points. For the LEED BD+C rating systems, these points are:

a) 9 points from Project Priorities credit and 1 point from LEED AP credit

b) 6 points from Project Priorities credit and 4 point from LEED AP credit

c) 8 points from Project Priorities credit, 1 point from LEED AP credit, and 1 point from Innovation

d) 9 points from Project Priorities credit and 1 point from Regional Priority

83) An energy analyst is selecting which ASHRAE 90.1-2022, Section 11.5.2 measures may be counted toward incremental points under Enhanced Energy Efficiency, Option 1, Path 1. Which of the following measures is eligible to contribute toward points under this path?

a) Building mass with night flush

b) On-site renewable energy generation

c) Improved envelope performance

d) Energy monitoring

84) A public elementary school will also serve as a community heating and cooling center and emergency meeting point, and the owner sets a 100-year design service life. The team is selecting the emissions scenario for the Climate Resilience Assessment prerequisite. Which approach is consistent with LEED's guidance?

a) Select a lower-warming scenario, because designing against severe conditions would over-engineer the building and waste embodied carbon

b) Select the scenario that matches current observed conditions, because projections beyond a few decades are not actionable

c) Select a higher-warming scenario, because a building that must stand for many decades will face the more severe climate conditions expected later in the century

d) Scenario selection is independent of service life; the same scenario applies to every project in a given region

85) A team is selecting from the nine product categories eligible under the Building Product Selection and Procurement credit. Which of the following product categories is eligible under this credit but is not one of the categories assessed under the Low-Emitting Materials credit?

a) Composite wood

b) Adhesives and sealants

c) Plumbing fixtures

d) Insulation

86) A qualifying transit route provides weekend service of 130 trips on Saturdays and 95 trips on Sundays. According to the primary requirements for the public transit service path of the Compact and Connected Development credit, which weekend trip count must the project use toward the threshold?

a) 130 trips, the higher weekend day

b) 95 trips, the lower weekend day

c) 112 trips, the two-day weekend average

d) 225 trips, the combined weekend total

87) To reduce air quality problems through scheduling under the construction management prerequisite, a project sequences its interior finishes. According to the prerequisite's best-practice scheduling guidance, when should absorptive-finish materials such as carpet and ceiling tile be installed relative to wet-applied materials?

a) Concurrently with paints and stains to compress the construction schedule

b) After wet-applied materials such as paints and stains have fully cured and are completely dry

c) Before paints and stains so the absorptive materials are protected by later coatings

d) Immediately after the first coat of wet-applied material, before it has fully cured

88) A project team pursuing the no-irrigation path of the outdoor water use option under Enhanced Water Efficiency plans to use temporary irrigation while new plantings establish. What is the maximum establishment period during which temporary irrigation is permitted?

a) One year

b) Two years

c) Three years

d) No temporary irrigation is permitted under this path

89) A project with demolition in scope conducts the salvage assessment required by the materials reuse option of Building and Materials Reuse. How do the salvaged materials contribute to LEED credit, depending on their destination?

a) Materials retained on-site contribute to Building and Materials Reuse, while salvaged materials sent for off-site reuse contribute to Construction and Demolition Waste Diversion

b) All salvaged materials contribute to Construction and Demolition Waste Diversion regardless of destination

c) Materials sent off-site contribute to both credits simultaneously

d) All salvaged materials contribute to Building and Materials Reuse regardless of destination

90) A New Construction laboratory near a region prone to wildfires wants its HVAC systems to switch into a smoke-ready operating mode during periods of heightened outdoor air pollution. Which option of the Resilient Spaces credit corresponds to this strategy, and what standard governs it?

a) Management Mode for Episodic Outdoor Ambient Conditions, governed by ASHRAE Guideline 44

b) Management Mode for Respiratory Diseases, governed by ASHRAE Standard 241-2023

c) Design for Occupant Thermal Safety During Power Outages, governed by ASHRAE Guideline 44

d) Management Mode for Episodic Outdoor Ambient Conditions, governed by ASHRAE Standard 241-2023

91) For a project with incomplete space, the owner must attest to something specific. What does the required letter of commitment state?

a) That the incomplete spaces will remain unoccupied until fit-out

b) That the incomplete spaces will be leased within five years

c) That the incomplete spaces will pursue LEED ID+C certification

d) That the incomplete spaces will satisfy the requirements of each prerequisite and credit achieved by the project

92) A property manager asks whether e-cigarettes and cannabis fall under the smoking prohibitions of No Smoking or Vehicle Idling. What is the prerequisite's scope?

a) Yes; the prerequisite minimizes exposure to tobacco smoke and smoke from tobacco substitutes or cannabis, and smoking prohibitions extend to electronic smoking devices
b) No; only combustible tobacco is covered
c) Cannabis is covered only where it is illegal
d) E-cigarettes are covered only indoors

93) A team is preparing the site energy estimate required by the Operational Carbon Projection and Decarbonization Plan prerequisite. What does site energy represent?

a) The energy measured at the building, before any losses from generating or delivering it
b) The energy consumed at the power plant to serve the building, including transmission losses
c) The building's energy use after applying the source energy conversion factors
d) The building's peak demand across the highest-consuming month of the year

94) When a product is tested with a qualifying independent laboratory report rather than a third-party certification under the low-emitting materials credit, the VOC emissions evaluation references a specific test method and modeling scenario. Which combination is required?

a) ANSI/BIFMA M7.1-2011 (R2021), using the open plan scenario
b) CDPH Standard Method v1.1-2010, using the private office scenario
c) CDPH Standard Method v1.2-2017, using the private office scenario
d) SCAQMD Rule 1168, using the schools scenario

95) A zero lot line project pursues the previously-disturbed-sites path of the preserve-and-restore option. Because of its constrained footprint, the Biodiverse Habitat credit applies reduced restoration thresholds. What percentage of previously disturbed area must a zero lot line project restore to earn one point?

a) 10% of the previously disturbed area
b) 20% of the previously disturbed area
c) 5% of the previously disturbed area
d) 40% of the previously disturbed area

96) Which of the following is the vision statement of USGBC?

a) Buildings and communities will regenerate and sustain the health and vitality of all life within a generation
b) Establishing leadership, promoting the triple bottom line, maintaining integrity, and fostering social equity
c) Recognize environmental leadership in the building industry and stimulate green competition
d) To transform the way buildings and communities are designed, built, and operated, enabling an environmentally and socially responsible, healthy, and prosperous environment that improves the quality of life

97) Under the Renewable Energy credit, what must happen to the renewable energy certificates or energy attribute certificates associated with the project's renewable energy?

a) They must be sold on the open market to offset the system's cost
b) They must be retired on behalf of the LEED project to prevent double counting
c) They may be retained by the utility that interconnects the system
d) They must be transferred to USGBC for tracking

98) A project located in a hurricane-prone area addresses hurricanes under the Enhanced Resilient Site Design credit. Which standard applies?

a) The FORTIFIED Commercial Wind standards
b) The FORTIFIED Commercial High Wind and Hail Specific Design Requirements
c) FEMA 543
d) The U.S. National Tsunami Hazard Mitigation Program

99) How many interest groups are represented on LEED consensus committees?

a) 3
b) 5
c) 7
d) 9

100) A mechanical engineer employed by the same firm that designed the building's HVAC systems is proposed as the commissioning provider (CxP) under Fundamental Commissioning. Under what condition is this arrangement acceptable?

a) Never; the CxP must always be a third-party entity holding no other contract on the project
b) If the engineer is a completely independent member of the design team, not directly associated with any aspect of the design or installation of the building systems
c) If the owner co-signs each commissioning report prepared by the engineer
d) If the engineer's commissioning scope is limited to the building envelope

PRACTICE TEST 3 – ANSWERS

A score above 80 would indicate very well preparation for the exam.

1	D	**26**	A	**51**	D	**76**	D
2	B	**27**	A	**52**	A	**77**	C
3	A	**28**	B	**53**	A	**78**	B
4	A	**29**	A	**54**	A	**79**	B
5	B	**30**	A	**55**	A	**80**	A
6	C	**31**	C	**56**	A	**81**	D
7	B	**32**	B	**57**	C	**82**	A
8	B	**33**	C	**58**	B	**83**	A
9	D	**34**	A	**59**	A	**84**	C
10	C	**35**	B	**60**	A	**85**	C
11	B	**36**	D	**61**	C	**86**	A
12	C	**37**	C	**62**	B	**87**	B
13	D	**38**	D	**63**	D	**88**	B
14	D	**39**	B	**64**	C	**89**	A
15	C	**40**	A	**65**	B	**90**	A
16	C	**41**	C	**66**	B	**91**	D
17	D	**42**	C	**67**	B	**92**	A
18	D	**43**	B	**68**	A	**93**	A
19	B	**44**	C	**69**	C	**94**	C
20	C	**45**	C	**70**	B	**95**	A
21	D	**46**	C	**71**	A	**96**	A
22	C	**47**	B	**72**	A	**97**	B
23	A	**48**	C	**73**	B	**98**	A
24	A	**49**	B	**74**	C	**99**	C
25	D	**50**	A	**75**	C	**100**	B

PRACTICE TEST 3 ANSWERS & EXPLANATIONS

1) Correct answer: D. The 70th, 75th, and 80th percentile events, in place of the standard 80th, 85th, and 90th.

Under Rainwater Management, zero lot line projects use reduced percentile tiers: the 70th, 75th, and 80th percentile regional or local rainfall events, where standard projects manage the 80th, 85th, and 90th. A building occupying its entire parcel has little room for infiltration and bioretention, so each tier drops. Zero lot line projects may also combine on-site and off-site strategies coordinated with local or regional authorities — a flexibility no other project type receives.

2) Correct answer: B. Solid glass enclosures or doors that seal when closed.

Under Fundamental Air Quality, indoor fireplaces and woodstoves must have solid glass enclosures or doors that seal when closed. Units that are not closed combustion or power-vented must additionally pass a backdraft potential test demonstrating that depressurization of the combustion appliance zone stays below 5 pascals. CO monitors are required throughout dwelling units regardless — one per floor, hard-wired with battery backup.

3) Correct answer: A. The environmental benefits associated with one megawatt-hour of renewable electricity generation.

A renewable energy certificate (REC) — like a guarantee of origin — corresponds to the environmental benefits of one megawatt-hour of renewable electricity generation. The certificate is the tradable proof of the clean attributes of that energy, distinct from the electricity itself. It measures generated energy (megawatt-hours), not installed capacity (kilowatts), a span of operating time, or a percentage of the project's own consumption.

4) Correct answer: A. ASHRAE Guideline 42-2023.

Both options of the credit use measures described in ASHRAE Guideline 42-2023, the design guide for enhanced indoor air quality. The other standards address ventilation rates, filtration, and air cleaning rather than the enhanced-IAQ design measures.

5) Correct answer: B. Building systems that condition spaces and provide comfort and basic amenities, such as interior and exterior lighting, HVAC, hot water heating, toilet exhaust, and parking garage ventilation.

Nonprocess energy, also called regulated energy, covers the systems that condition spaces and provide occupant comfort and basic amenities — interior and exterior lighting, HVAC, service hot water heating, toilet exhaust, and parking garage ventilation. These are the loads the ASHRAE 90.1 prescriptive sections govern, which is why the credit's regulated-loads path is built on them. The equipment tied to building use rather than comfort describes process (unregulated) energy, and neither renewable generation nor generation-and-delivery losses defines regulated energy.

6) Correct answer: C. Asphalt, concrete, masonry, structural steel, insulation, aluminum extrusions, structural wood and composites, cladding, and glass.

The prerequisite states that at a minimum the quantification must include asphalt, concrete, masonry, structural steel, insulation, aluminum extrusions, structural wood and composites, cladding, and glass. It is not limited to only the highest-impact materials, and mechanical, electrical, plumbing, and conveying equipment are not part of this required minimum list.

7) Correct answer: B. On-street parking located in public rights-of-way.

The path specifies that parking capacity must include all existing and new off-street parking spaces that are leased or owned by the project, including parking outside the project boundary but used by the project. The single explicit carve-out is that on-street parking in public rights-of-way is excluded from the calculation.

8) Correct answer: B. That operator training met the training plan and actually took place.

In both enhanced commissioning paths, the commissioning provider must verify that operator training met the training plan and actually occurred. Confirming both the content (against the plan) and the fact of delivery ensures the people running the building can operate the commissioned systems as intended — commissioning gains are quickly lost if operators are never properly trained.

9) Correct answer: D. Views must be through glass with visible light transmittance above 40%, and where glazing has frits, patterns, or tints, the view must be preserved.

Under Occupant Experience, view glazing must have visible light transmittance above 40%. Frits, patterns, and tints do not automatically disqualify the glass — the view must be preserved, and neutral gray, bronze, and blue-green tints are acceptable.

10) Correct answer: C. The credit is not available, because the Green Leases credit applies only to Core and Shell projects.

The Green Leases credit is available only to LEED BD+C: Core and Shell projects. Its intent is to ensure tenants complete the sustainable design and construction features started by the base building, which is a Core and Shell condition — the developer delivers the shell and the tenant performs the fit-out. New Construction projects control their own fit-out, so the credit does not apply regardless of how many tenants occupy the building.

11) Correct answer: B. It counts as a nonroof measure under the Heat Island Reduction credit by providing shade, and separately supports the Renewable Energy credit by generating on-site power.

The Heat Island Reduction credit lists structures covered by energy generation systems, such as solar thermal collectors, photovoltaics, and wind turbines, among its nonroof measures — the structure earns credit for the shade it casts over pavement. The same array simultaneously produces on-site clean power that supports the Renewable Energy credit.

12) Correct answer: C. That it be installed using native or adapted plant species.

The vegetated roof measure strategy requires installing a vegetated roof using native or adapted plant species.

13) Correct answer: D. Salvage experts such as deconstruction contractors who know how to safely remove materials and which materials carry salvage value.

Under Building and Materials Reuse, successful salvage assessments involve salvage experts, such as deconstruction contractors, who know how to safely remove building materials and which materials have salvage value. Identifying destinations during the assessment increases success — contacting salvage retailers and verifying they will accept or purchase a particular material saves time and resources, and some organizations offer pickup depending on quantity, distance, and value.

14) Correct answer: D. A CO monitor on each floor of each dwelling unit, hard-wired with battery backup, in all unit types regardless of the equipment installed.

Under Fundamental Air Quality, each dwelling unit needs a carbon monoxide monitor on each floor. Monitors must be hard-wired with battery backup. The requirement applies to all unit types, regardless of installed equipment. Even an all-electric unit needs monitors, because CO can migrate from garages, neighboring units, or outdoors.

15) Correct answer: C. Decorative lighting for a private rooftop terrace used for evening events.

The exemptions cover specialized signal, directional, and marker lighting for transportation; facade and landscape lighting in lighting zones 3 and 4 that is automatically turned off from midnight until 6 a.m.; lighting for theatrical purposes for stage, film, and video performances; government-mandated roadway lighting; lighting for hospital emergency departments, including associated helipads; lighting for the national flag in lighting zones 2, 3, or 4; and internally illuminated exterior signage.

16) Correct answer: C. 25 mph (40 km/h).

The bicycle network may consist of bicycle paths, lanes, multiuse trails, or streets with a maximum speed limit of 25 mph (40 km/h).

17) Correct answer: D. A minimum of 15 years, with existing dwelling units excluded from the calculations.

Rental units must be maintained at affordable levels for a minimum of 15 years, and existing dwelling units are exempt from the requirement calculations so that only new affordable units count.

18) Correct answer: D. Planning for Zero Waste Operations.

Under Option 1, tenant requirements must ensure compliance with nine prerequisites for any tenant-installed systems: Climate Resilience Assessment, Human Impact Assessment, Carbon Assessment, Minimum Water Efficiency, Minimum Energy Efficiency, Fundamental Commissioning, Energy Metering and Reporting, Planning for Zero Waste Operations, and Fundamental Air Quality.

19) Correct answer: B. Path 1 caps peak load intensity using the Passive House tools, WUFI Passive or the Passive House Planning Package.

Option 4 first requires measured air leakage, balanced ventilation, and inclusion of ventilation loads in the calculations, then offers three ways to quantify the reduction. Path 1, Peak Load Intensity, caps the peak load intensity using the Passive House tools — the WUFI Passive design tool or the Passive House Planning Package. Path 2 demonstrates percentage improvements through the ASHRAE 90.1-2022 envelope trade-off analysis and/or the Total System Performance Ratio method. Path 3, Energy Simulation, calculates a performance index under ASHRAE 90.1 Appendix G.

20) Correct answer: C. Base ratios from the Institute of Transportation Engineers Parking Generation Manual, sixth edition, or a comparable resource applied by a qualified transportation engineer or planner, or supplementary LEED guidance.

Under Transportation Demand Management, the reduce parking path measures reductions against base ratios from the Institute of Transportation Engineers Parking Generation Manual, sixth edition, a comparable resource applied by a qualified transportation engineer or planner, or supplementary LEED guidance.

21) Correct answer: D. LEED BD+C: Core and Shell, because at least 40% of the floor area is incomplete.

LEED BD+C: Core and Shell is required when at least 40% of the project's gross floor area is incomplete at certification — meaning it lacks basic finishes or essential MEP systems. With 48% incomplete, this project exceeds the 40% threshold and cannot use New Construction (which requires at least 60% complete). O+M applies to buildings operational for at least one year. ID+C applies to interior renovations with no exterior scope.

22) Correct answer: C. 100-year GWPs assessed in accordance with the IPCC Fourth Assessment Report or later.

The prerequisite references 100-year GWPs assessed in accordance with the IPCC Fourth Assessment Report or later.

23) Correct answer: A. Construction and Demolition Waste Diversion.

Materials salvaged and sent off-site to reuse markets count toward the Construction and Demolition Waste Diversion credit, since they leave the project as diverted waste.

24) Correct answer: A. A description of the regional or local priority, with evidence and description of strategies adopted by the project to address it.

For regional priority strategies, the team must also include a description of regional/local priority, with evidence and description of strategies adopted by the project to address the aforementioned priority. The pilot-registration evidence is for pilot credits, the measurable-improvement summary is for innovative strategies not in the library, and the LEED AP confirmation belongs to a separate credit entirely, not to the regional priority documentation.

25) Correct answer: D. Tier 2 must be operational within two years of occupancy or have a commercial operations date no more than five years before the contract, while Tier 3 electricity may be up to fifteen years old.

The age of the renewable generator is the key distinction. Tier 2 (New Off-site Renewable Electricity) requires electricity contracted to be operational within two years of building occupancy, or generated by an asset with a commercial operations date (COD) no more than five years before contract execution. Tier 3 (Off-site Renewable Energy), by contrast, may have a COD up to fifteen years old to meet the Green-e Energy generator-age criteria.

26) Correct answer: A. Case 3 uses the ASHRAE 90.1 exceptional calculation method to demonstrate a percentage improvement in plug, process, refrigeration, and conveyance loads; Cases 1 and 2 may be combined, but Case 3 is pursued on its own.

Case 3 applies the ASHRAE 90.1 exceptional calculation method to demonstrate a percentage improvement in plug, process, refrigeration, and conveyance loads — a modeling-based route to the same credit. On combination, Cases 1 and 2 may be combined with each other, but Case 3 is pursued on its own, because its whole-set calculation would otherwise double-count the operational and equipment measures. Case 3 is not limited to refrigeration, does not require the dashboard, and the cases are not all mandatory together.

27) Correct answer: A. Equitable Development adds the sensitive project location requirement, and Compact and Connected Development adds the surrounding development and resources option.

For data centers, warehouses, and distribution centers, two LT credits carry bespoke content. Equitable Development requires the standard options plus the sensitive project location requirement — the building at least 300 feet (90 meters) from sensitive receptors. Compact and Connected Development offers the surrounding development and resources option, with its development-and-adjacency and transportation-resources paths. Sensitive Land Protection, TDM, and Electric Vehicles treat logistics projects like any other.

28) Correct answer: B. Teams running an energy model for the Minimum Energy Efficiency prerequisite may take the values from the simulation results; teams relying on similar projects or published data may estimate from predicted design conditions.

The prerequisite accommodates both situations. A team already performing a simulation for the Minimum Energy Efficiency prerequisite simply generates the required data from the simulation results — no duplicate modeling. A team working from similar projects or published data instead establishes estimates based on the predicted design conditions, including designed system characteristics. Requiring a calibrated model of every project would defeat the accessibility this prerequisite is built around, and the estimate is a design-phase projection, so it cannot depend on metered operating data that does not yet exist.

29) Correct answer: A. Adapted from OSHA heat-prevention guidance; measures include providing cool, shaded, or air-conditioned rest areas, required rest breaks, and scheduling labor-intensive work for cooler parts of the day.

Under Construction Management, the extreme heat requirement is adapted from OSHA prevention guidance for preventing heat-related illness, not from SMACNA (which is the basis for the air quality practices), ASHRAE 55, or the WHO guidelines. Preventive measures include providing cool, shaded, or air-conditioned areas for rest, implementing required rest breaks, and scheduling labor-intensive activities in cooler parts of the day.

30) Correct answer: A. 100 feet (30 meters) from the water body and 50 feet (15 meters) from the wetland.

Under the previously undeveloped sites option, the development footprint cannot be on or within 100 feet (30 meters) of a water body or on or within 50 feet (15 meters) of a wetland, except for minor improvements that allow human interaction without significantly altering the existing vegetation and hydrology.

31) Correct answer: C. At least 10% of peak electricity demand for a minimum of one hour.

Under the building automation system path, the team must develop a plan for shedding at least 10% of the project's peak electricity demand for a minimum of one hour, addressing both winter and summer peaks while considering electrified grid projections.

32) Correct answer: B. Calculated luminance of less than 6,000 cd/sq. m. between 45 and 90 degrees from nadir, and a color rendering index of at least 90.

Under Occupant Experience, Path 2 (Quality Electric Lighting), electric light glare control is met when each luminaire has a calculated luminance of less than 6,000 candela per square meter (cd/sq. m.) between 45 and 90 degrees from nadir, or achieves a UGR of 19 or lower (not 25). For color rendering, luminaires must have a color rendering index of at least 90, or meet the IES TM-30-20 requirements (Rf of 78 or higher, Rg of 95 or higher, Rcs,h1 of –1% to 15%).

33) Correct answer: C. Divert at least 50% of total material, with at least 10% of diverted materials salvaged or source-separated to single-material recyclers.

For one point, the project must divert at least 50% of the total construction and demolition material, and at least 10% of the diverted materials must be salvaged or source-separated and sent to single-material recycler(s). The 75%/25% pairing earns two points, not one, so it is the wrong tier here.

34) Correct answer: A. 25%.

The two-point threshold requires diverting at least 75% of the total construction and demolition material, with at least 25% of the total diverted materials salvaged or source-separated and sent to single-material recycler(s). With 78% diverted, the team clears the overall bar, but at least 25% of those diverted materials must come from salvage or source separation.

35) Correct answer: B. No; composite wood uses the formaldehyde emissions evaluation criteria, such as ULEF or NAF certification under EPA TSCA Title VI or CARB ATCM, not the CDPH method.

Under Low-Emitting Materials, composite wood products must meet the formaldehyde emissions evaluation criteria. Compliance routes include certification as ultra-low-emitting formaldehyde (ULEF) or no added formaldehyde (NAF) under EPA TSCA Title VI or the CARB Airborne Toxic Control Measure, from a CARB-approved or EPA-recognized third-party certifier. Wood structural panels labeled Exposure 1 or Exterior under PS 1 or PS 2 also qualify. So do structural wood products made to the listed ASTM and ANSI standards, and inherently non-emitting, salvaged, or reused products. The CDPH Standard Method v1.2-2017 is the real basis of the VOC emissions evaluation criteria used by the other product categories. Composite wood runs on the formaldehyde track instead.

36) Correct answer: D. 10 spaces.

The single point available under the readiness option requires commercial projects to provide EV-ready spaces for the greater of 10% of total parking or 10 spaces. For 80 spaces, 80 x 0.10 = 8 spaces, but the 10-space floor governs because it is greater, so the project must design 10 EV-ready spaces.

37) Correct answer: C. A cost-benefit analysis comparing the proposed resilience strategies against projected avoided losses.

For each priority hazard the team must assess and specify the Intergovernmental Panel on Climate Change emissions scenario used, specifying the shared socioeconomic pathways; the projected service life of the project; the hazard level; the hazard risk rating; exposure, sensitivity, adaptive capacity, vulnerability, and overall risk levels; the potential impact on the project site and building function; and the potential impact on the project site during construction.

38) Correct answer: D. Opportunity for daylight, views, and operable windows; environmental conditions of air and water; and adjacent soundscapes, lighting, and wind patterns.

Under Human Impact Assessment, the occupant experience category covers three groups of factors. The first is an opportunity for daylight, views, and operable windows. The second is the environmental conditions of air and water. The third is adjacent soundscapes, lighting, and wind patterns within the context of surrounding buildings, such as a microclimate or solar scape.

39) Correct answer: B. A code-minimum version of the same building.

The Energy Cost Budget method uses whole-building energy modeling in which the team models the proposed design and demonstrates it performs at least as well as a code-minimum version of the same building. The budget building therefore tracks the proposed design's own form and configuration. This is the crucial contrast with the Appendix G Performance Rating Method, which compares the design against a fixed baseline building that does not shift with the design.

40) Correct answer: A. The project's EPA Level III ecoregion, or a local equivalent for projects outside the U.S.

The prerequisite requires that planted vegetation be native or adapted to the project's EPA Level III ecoregion or local equivalent for projects outside of the U.S., and that no planted vegetation be classified as invasive at the time of installation.

41) Correct answer: C. 36 months.

The monitoring system must be capable of reporting at least hourly, daily, monthly, and annually, and must maintain all data collected for 36 months.

42) Correct answer: C. Gold, because the embodied carbon reduction falls below the required 20% threshold for Platinum.

LEED v5 Platinum requires both 80+ points AND meeting all four decarbonization thresholds: electrification (5 points), enhanced energy efficiency (8 points), 100% renewable energy, and 20% embodied carbon reduction. This project meets three thresholds but falls short on embodied carbon (15% vs. 20% required). Since all four must be met, the project cannot achieve Platinum and defaults to Gold.

43) Correct answer: B. The meters need not be utility owned or utility grade.

For residential projects, a permanent water meter measuring each dwelling unit's total potable water use is installed, and these meters need not be utility owned or utility grade. A single whole-building meter does not satisfy the per-unit residential requirement of the credit.

44) Correct answer: C. It is a supporting resource that supplies practical checklists and agendas for facilitating well-structured meetings, workshops, and charrettes.

The Integrative Design Process credit requires teams to implement equitable processes by facilitating well-structured meetings, workshops, and charrettes, and names the U.S. Department of Energy's Handbook for Planning and Conducting Charrettes for High-Performance Projects as a resource that provides practical checklists and agendas to support that work. It is a supporting resource, not a referenced standard.

45) Correct answer: C. 80%.

Path 2, Low GWP Refrigerants, caps the total weighted average refrigerant GWP across all new equipment at 80% of the weighted average benchmarks for the first point; a deeper reduction to 50% earns the second. 70% and 90% are near-miss values not used.

46) Correct answer: C. At least 25% of the required outdoor space, planted with two or more types of vegetation or with an overhead vegetative canopy.

At least 25% of the required outdoor space must be vegetated and planted with two or more types of vegetation or have an overhead vegetative canopy.

47) Correct answer: B. Projects must evaluate the grid-interactive measures against the current and forecasted grid context, and provide interval recording meters and equipment capable of accepting an external signal.

For all options, the design must evaluate the grid-interactive measures against the current and forecasted grid context and include interval recording meters plus equipment capable of accepting an external signal indicating when load management should be enabled. These common elements make any grid-interactive strategy measurable and responsive to grid conditions. On-site storage, demand response enrollment, and 72-hour islanding each belong to a specific option, not to every option.

48) Correct answer: C. As a multi-attribute product.

The credit states that products that achieve two or more criteria areas are considered multi-attribute, and such products earn a higher value in credit calculations.

49) Correct answer: B. The default gender mix is half male and half female, and modifications must be shown to apply for the life of the building.

The default gender mix is half male and half female. Modifications to the 50:50 ratio must be shown to apply for the life of the building, such as a single-gender educational facility, and must be supported by a narrative and data.

50) Correct answer: A. Natural Resources Conservation Service, NatureServe, and the U.S. Endangered Species Act.

The referenced standards for the previously undeveloped sites option are the Natural Resources Conservation Service, NatureServe, and the U.S. Endangered Species Act. These support verification of prime farmland, imperiled species communities, and threatened or endangered species, respectively.

51) Correct answer: D. From USGBC's Project Priority Library.

Under the regional priority pathway, the team must achieve a regional priority credit from USGBC's Project Priority Library; these credits have been identified by USGBC as having additional regional importance for the project's region. The same Project Priority Library is also the source for the project-type, exemplary performance, and pilot credit pathways.

52) Correct answer: A. 1 point.

The LEED AP credit awards a single point when at least one principal team member holds a LEED AP with a BD+C specialty. Having three LEED APs does not multiply the award — the credit is worth 1 point regardless of how many qualified professionals are on the team.

53) Correct answer: A. Construction and Demolition Waste Diversion.

The Construction and Demolition Waste Diversion credit covers construction-phase waste, while this prerequisite covers operations-phase waste. The other credits address reuse, embodied carbon, and product selection rather than construction waste diversion.

54) Correct answer: A. Industry average values.

Path 1, the Project-Average Approach, compares the whole-project weighted average embodied carbon of the procured materials against industry average values, using a tool such as EC3.

55) Correct answer: A. The operations phase.

This prerequisite addresses waste generated during building operations. Construction- and demolition-phase waste is handled by the Construction and Demolition Waste Diversion credit, so the two cover different phases of the building's life.

56) Correct answer: A. They automatically comply with the VOC emissions evaluation and do not require emissions evaluations.

Under Low-Emitting Materials, products that are salvaged and reused and more than one year old will automatically comply with the VOC emissions evaluation and do not require emissions evaluations. However, if a product with a separate manufacturer and cost is later applied to the salvaged item, that applied product is subject to the applicable low-emitting criteria; and for salvaged composite wood, any off-site applied finishes must meet the VOC criteria. The credit's standard VOC pathway uses the CDPH Standard Method v1.2-2017 private office scenario, but aged salvaged/reused materials bypass that testing.

57) Correct answer: C. A reduction of at least 50% in the Theoretical Irrigation Requirement.

Option 4 requires the project to reduce its Theoretical Irrigation Requirement by at least 50% from the calculated baseline, achieved through plant species selection, irrigation system efficiency, or a combination. The 30% figure is the corresponding threshold in the Minimum Water Efficiency prerequisite's efficient irrigation option.

58) Correct answer: B. Seven.

Under Air Quality Testing and Monitoring, measurement counts scale with total occupied floor area. Between 25,000 and 200,000 square feet, the count is four plus one per each full 25,000 square feet above 25,000. At 120,000 square feet, the area above the threshold is 95,000 square feet. That yields three full increments. The total is 4 + 3 = 7. Only completed increments count, never rounded-up fractions. Above 200,000 square feet, the formula shifts to ten plus one per 50,000 square feet.

59) Correct answer: A. 40%.

Views must be through glazing with a visible light transmittance above 40%, so the view is clear rather than heavily tinted.

60) Correct answer: A. The team must update the analysis because the quantities changed more than 10% from design through construction.

The credit permits both baseline and final results to use as-designed or as-constructed quantities provided that quantities did not change more than 10% from design through construction. A 14%

increase breaches that limit, so the team must update its analysis (for both proposed and baseline) to accurately represent what was installed.

61) Correct answer: C. Visual, tactile, contrasting, or photoluminescent warnings at floor-level changes.

Under Accessibility and Inclusion, the safety and aging category includes providing visual, tactile, contrasting, or photoluminescent warnings at floor-level changes. Photoluminescent markings keep working during power loss, and tactile warnings serve users with low vision. The category also carries nonslip flooring, fixed area rugs with transition strips, visual indication or railings at full-height glazing, audible and visual emergency alerts, closed risers, and visual contrast between walls and floors, doors, and casework.

62) Correct answer: B. The LEED addenda process.

Under Building Product Selection and Procurement, the level definitions and the product documentation scoring table are defined by USGBC in the credit library resources. They will be updated over time through the LEED addenda process — the formal mechanism by which USGBC corrects and evolves rating system content between versions.

63) Correct answer: D. Up to 50 feet (15 meters) from grade, and up to 20 feet (six meters) from the finished grade of a green roof.

The bird-friendly glass option applies to all glass, including spandrel glass, located from grade up to 50 feet (15 meters) measured at all points and up to 20 feet (six meters) measured from the finished grade of a green roof, plus glass at any distance in guardrails and windshields.

64) Correct answer: C. An open and accessible route to and around the equipment.

Under Accessibility and Inclusion, the physical diversity category includes incorporating accessible and inclusive equipment and activities in fitness facilities — and ensuring an open and accessible route to and around the equipment. Accessible machines behind an impassable layout serve no one; circulation is half the strategy. The category's neighbors include touchless door operators, meeting spaces accommodating mobility devices for 10% of occupants, and alternate accessible routes paralleling stairs.

65) Correct answer: B. Carries out the functional performance testing and confirms that portions of the project meet their stated requirements.

A verification and testing provider is the entity that completes the activities needed to implement functional performance testing and confirms that parts of the project meet their stated requirements. For envelope scope specifically, that verification includes a design review, periodic field inspection, and reporting. The provider executes and confirms testing rather than authoring the owner's project requirements, which the commissioning provider assists the owner in developing.

66) Correct answer: B. More than 20%.

Option 3, GreenChill Certification for Food Retailers, is limited to projects where food retailing constitutes more than 20% of the gross area — the point at which large commercial refrigeration systems become significant.

67) Correct answer: B. Dedicated equipment space, chase ways, junction boxes and dedicated electrical panel capacity, and accommodations supporting energy recovery ventilation for at least 50% of the ventilation air, with the details documented in the tenant guidelines.

Electrification Readiness provides the building infrastructure that ensures the building can later operate without on-site combustion and can accept efficient electric heating and service water heating equipment. The required provisions, as applicable and appropriately sized, are dedicated physical space for future electric equipment, chase ways for refrigerant lines, condensate, and other piping, a junction box with dedicated electrical panel capacity for a properly phased branch circuit where electrical distribution is in scope, and, for portions of the building where ventilation air is not installed in the project scope, accommodations capable of supporting energy recovery ventilation for at least 50% of the ventilation air. The details must be captured in the project plans and the tenant guidelines, with guidance for designing efficient electrified systems. The option is deliberately about readiness, not installation — it prepares the building so a future tenant can electrify efficiently.

68) Correct answer: A. During predesign or very early in the design phase.

The Enhanced Commissioning credit requires the commissioning provider to be engaged during predesign or very early in the design phase, well before the Fundamental Commissioning prerequisite requires it. Early engagement lets the provider shape the owner's project requirements and catch design issues while changes are still inexpensive — the whole value of enhanced commissioning lies in this earlier, deeper involvement.

69) Correct answer: C. Passing spaces of 60 inches (1.525 meters) at least every 200 feet (61 meters) along any path less than 60 inches (1.525 meters) wide.

Under Accessibility and Inclusion, the clear width provisions require a minimum clear width of 32 inches (0.815 meters) at doorways meant for human passage and 36 inches (0.915 meters) for circulation paths. Along any path less than 60 inches (1.525 meters) wide, passing spaces of 60 inches (1.525 meters) must occur at least every 200 feet (61 meters). Teams are encouraged toward 36-inch doorways for enhanced accessibility. These baseline physical elements apply where the project's region has no accessibility code in place.

70) Correct answer: B. Every outdoor air fan system must include energy recovery at a 70% enthalpy or 75% sensible recovery ratio, with bypass controls for moderate conditions; systems supplying less than 15% of the total outdoor air may be excluded in aggregate.

Option 2 requires each outdoor air fan system to provide energy recovery at a minimum 70% enthalpy recovery ratio or a minimum 75% sensible heat recovery ratio, with provisions to bypass or control the energy recovery system during moderate outdoor air conditions so it does not impose a penalty when recovery is not beneficial.

71) Correct answer: A. U.S. EPA or the Carbon Leadership Forum Material Baselines.

The industry averages come from the U.S. EPA, the Carbon Leadership Forum (CLF) Material Baselines report, or similar recognized sources.

72) Correct answer: A. A representative from the base building team or the building's owner, responsible for coordinating communication about the base building's sustainable design and construction documentation.

The guidelines must designate a specific point of contact, either from the base building team or the building's owner, who is responsible for coordinating communication about the base building's sustainable design and construction documentation, providing tenants with resources and support, and delivering training sessions. The role belongs to the base building side, not to an independent commissioning agent, a USGBC reviewer, or a tenant's facility manager, so each of those alternatives misidentifies who holds the responsibility.

73) Correct answer: B. ENERGY STAR.

For residential and commercial clothes washers, residential dishwashers, and ice machines, teams in the U.S. and Canada must use ENERGY STAR-labeled equipment; for international projects a performance equivalent is accepted. The split is worth memorizing: WaterSense governs the fixture group — toilets, urinals, private lavatory faucets, and showerheads — while ENERGY STAR governs the appliance and equipment group.

74) Correct answer: C. The MLO lighting zone together with the luminaire's mounting location and its distance from the lighting boundary, measured in mounting heights.

Under Light Pollution Reduction, maximum backlight and glare ratings come from a matrix of the MLO lighting zone and the luminaire's mounting location and distance from the lighting boundary, measured in mounting heights.

75) Correct answer: C. During the preliminary review by GBCI.

No appeals can be made during the preliminary review, as the review is still not finalized.

76) Correct answer: D. Yes; teams may select any two options or paths, even those not identified as high priorities in the climate resilience assessment, though assessment findings should inform selection.

Under Resilient Spaces, teams leverage findings from the Climate Resilience Assessment and Human Impact Assessment to select strategies. But the credit expressly allows any two options or paths, even those not identified as a high priority. No amendment of the assessment is needed. Enhanced Resilient Site Design, by contrast, must address at least two of the highest priority hazards from the assessment. Site resilience is chained to the priority hazards; indoor resilient spaces are only informed by them.

77) Correct answer: C. Capable of recording at least hourly, with access in real time via local network, building management system, cloud service, app, or online database.

The submeters option requires that all submeters be capable of recording data at least hourly and that the facility manager and/or tenants be able to access the submeter data in real time via local network, building management system, cloud service, app, or online database.

78) Correct answer: B. The review fee is based on the project's gross floor area and membership status, while the registration fee is a flat fee.

The LEED registration fee is a flat fee paid when registering the project. The review (certification) fee varies based on the project's gross floor area and the organization's USGBC membership status — larger projects pay higher review fees. Both fees are paid through the LEED certification process, not to Prometric (which administers professional exams, not building certification).

79) Correct answer: B. A USGBC-approved data template or an approved third-party data source such as ENERGY STAR Portfolio Manager.

The prerequisite states that projects share data using a USGBC-approved data template or an approved third-party data source, such as ENERGY STAR Portfolio Manager.

80) Correct answer: A. Five.

Path 1 maps each primary space across five categories: noise exposure, acoustic comfort, acoustic privacy, communication, and soundscape management.

81) Correct answer: D. Views into interior atria may meet up to 30% of the required area.

Under Occupant Experience, the quality views path provides occupants a view to the outdoor natural or urban environment for 75% or 90% of all regularly occupied floor area. Views into interior atria may be used to meet up to 30% of the required area. Views must pass through glass with visible light transmittance above 40%, must include nature, urban landmarks, art, or objects at least 25 feet (7.5 meters) from the glazing exterior, and occupants must be within three times the head height of the glazing with direct access to the view. Auditoriums, dedicated videoconference rooms, and gymnasiums may be excluded.

82) Correct answer: A. 9 points from Project Priorities credit and 1 point from LEED AP credit.

LEED projects are scored out of 100 base points, with up to 10 bonus points available: 9 under the Project Priorities credit and 1 under the LEED AP credit.

83) Correct answer: A. Building mass with night flush.

The eligible measures for Path 1 are limited to HVAC, service water heating, lighting, and the single load management measure for building mass/night flush (G07).

84) Correct answer: C. Select a higher-warming scenario, because a building that must stand for many decades will face the more severe climate conditions expected later in the century.

The emissions scenario projects a level of warming for the 21st century, and the higher the warming level, the hotter the future the project is designed against. Because a building that must stand for many decades will encounter the more severe conditions expected later in the century, teams should match the scenario to the project's service life and design longer-lasting buildings against a higher-warming scenario.

85) Correct answer: C. Plumbing fixtures.

The nine product categories under this credit are paints and coatings, adhesives and sealants, flooring, walls, ceilings, insulation, furniture, composite wood, and plumbing fixtures. Plumbing fixtures (water closets, urinals, lavatory and kitchen faucets, and showerheads) appear here but are not among the Low-Emitting Materials product categories, which expressly treat plumbing as not applicable. Composite wood, adhesives and sealants, and insulation are all shared by both credits, so they are not unique to this credit.

86) Correct answer: A. 130 trips, the higher weekend day.

The requirements direct that for weekend trips, the project count only the weekend day with the highest number of trips, which is the 130 trips on Saturdays. This differs sharply from the weekday rule, which uses the lowest day.

87) Correct answer: B. After wet-applied materials such as paints and stains have fully cured and are completely dry.

The scheduling guidance advises installing absorptive-finish materials after wet-applied materials have fully cured—for example, installing carpet and ceiling tile after paints and stains are completely dry—so the porous finishes do not adsorb VOCs off-gassing from the curing coatings. Installing them concurrently, beforehand, or after only a single uncured coat would expose the absorptive materials to emissions.

88) Correct answer: B. Two years.

Under Enhanced Water Efficiency, the no-irrigation path requires showing the landscape needs no permanent irrigation system beyond a maximum two-year establishment period.

89) Correct answer: A. Materials retained on-site contribute to Building and Materials Reuse, while salvaged materials sent for off-site reuse contribute to Construction and Demolition Waste Diversion.

Under Building and Materials Reuse, projects with demolition in scope conduct the salvage assessment before deconstruction or demolition begins. The destination then splits the credit. Materials retained on-site count here, under the materials reuse option. Materials sent off-site for reuse count under Construction and Demolition Waste Diversion, where salvage is valued at twice the rate of other diverted materials. Nothing counts twice.

90) Correct answer: A. Management Mode for Episodic Outdoor Ambient Conditions, governed by ASHRAE Guideline 44.

Designing systems to operate a smoke-ready or episodic outdoor event management mode for events such as wildfire smoke is the Management Mode for Episodic Outdoor Ambient Conditions option, which is described by and governed by ASHRAE Guideline 44 on protecting building occupants from smoke during wildfire and prescribed burn events. ASHRAE Standard 241-2023 governs infectious aerosols under the separate Management Mode for Respiratory Diseases option, not smoke events.

91) Correct answer: D. That the incomplete spaces will satisfy the requirements of each prerequisite and credit achieved by the project.

Under the projects with incomplete space provisions, the owner provides a letter of commitment attesting that the incomplete spaces will satisfy the requirements of each prerequisite and credit achieved by the project. The certification claims the whole building, so the owner vouches that future fit-outs will not undo it. The team also describes the incomplete space, the remaining scope, the responsible entity, and the reason for incompleteness — and contacts USGBC where a credit adaptation is unclear.

92) Correct answer: A. Yes; the prerequisite minimizes exposure to tobacco smoke and smoke from tobacco substitutes or cannabis, and smoking prohibitions extend to electronic smoking devices.

Under No Smoking or Vehicle Idling, the stated intent is minimizing exposure to tobacco smoke, smoke from tobacco substitutes or cannabis, and vehicle emissions — and the smoking scope includes conventional cigarettes, cigars, and pipes, cannabis whether medical or recreational, and electronic smoking devices. The construction-phase prohibition under Construction Management carries the same breadth.

93) Correct answer: A. The energy measured at the building, before any losses from generating or delivering it.

Site energy is the energy measured right at the building, before any of the losses incurred in generating or delivering it — it is what the meter sees. This is deliberately distinguished from source energy, which accounts for the upstream generation and transmission losses and is reached by applying source energy conversion factors, the approach used in the energy efficiency requirements rather than here. Because this prerequisite asks teams to report consumption from each energy source, the site basis keeps the reporting tied to measurable delivered quantities.

94) Correct answer: C. CDPH Standard Method v1.2-2017, using the private office scenario.

The VOC emissions evaluation criteria require testing according to the California Department of Public Health (CDPH) Standard Method v1.2-2017, meeting the VOC limits in Table 4-1 of the private office scenario (products used in classrooms may instead be modeled using the schools or private office scenario). ANSI/BIFMA M7.1 is the test method for the separate furniture emissions evaluation, not the general VOC criteria. SCAQMD Rule 1168 defines adhesives and sealants but is not the VOC emissions test method here. For independent lab reports, the test must also have been performed no more than three years prior to purchase.

95) Correct answer: A. 10% of the previously disturbed area.

The points table sets zero lot line thresholds at 10% of previously disturbed area for 1 point and 20% for 2 points, which are half of the standard 20% and 40% tiers.

96) Correct answer: A. Buildings and communities will regenerate and sustain the health and vitality of all life within a generation.

"Buildings and communities will regenerate and sustain the health and vitality of all life within a generation" is the vision statement of USGBC.

And USGBC's mission is "to transform the way buildings and communities are designed, built, and operated, enabling an environmentally and socially responsible, healthy, and prosperous environment that improves the quality of life."

97) Correct answer: B. They must be retired on behalf of the LEED project to prevent double counting.

All environmental attributes — energy attribute certificates or renewable energy certificates — associated with the project's renewable energy must be retired on behalf of the LEED project. Retirement removes the certificate from circulation so the clean attributes can be claimed only once, preventing the same megawatt-hour from being counted by both the LEED project and another party.

98) Correct answer: A. The FORTIFIED Commercial Wind standards.

For projects in hurricane-prone areas, the site and site structures must be constructed according to the FORTIFIED Commercial Wind standards, or a local equivalent, applied together with federal, state, and local codes. The paired FORTIFIED distractor — the High Wind and Hail Specific Design Requirements — is the hail standard, and confusing the two is the most common error since both carry the FORTIFIED name. FEMA 543 does address flooding and high winds, but the credit cites it for the flooding hazard alongside ASCE 24.

99) Correct answer: C. 7.

LEED consensus committees represent seven interest groups: building owners and developers; architects, engineers, and designers; contractors and facility operators; product manufacturers; building occupants and community advocates; regulation and standards organizations; and utilities, academia, finance, and others. No single group may exceed 25% of voting membership.

100) Correct answer: B. If the engineer is a completely independent member of the design team, not directly associated with any aspect of the design or installation of the building systems.

Under Fundamental Commissioning, the CxP must be completely independent of the design and construction work, per the referenced ASHRAE 90.1 provisions. Independence does not require a separate firm. The CxP can be employed by the company performing the design or construction of the building. The condition is that the person acts as a completely independent member of the design team, not directly associated with any aspect of the design or installation of the building systems. The owner may also employ the CxP directly. A third-party entity is one acceptable route, not the only one. If the CxP's company has direct influence on the design and construction, the commissioning plan must address potential conflicts of interest.

FINAL EXAM TIPS

FINAL EXAM TIPS

- **Read carefully. Make sure you read all the questions and the choices very carefully. If a question seems to have more than one answer, make sure you thoroughly understand the question and pay special attention to the wording. It doesn't matter how well you know the exam content; if you don't read carefully, the exam can easily trick you into selecting the wrong answer.**
- **Passing score.** The LEED AP BD+C exam is scored between 125 and 200. A score of **170 or higher is required to pass**. However, the exam contains both scored and unscored questions. Unscored questions are used by GBCI to collect performance data and potentially serve as future exam questions. Test-takers will not know which questions are scored and which are unscored. Because of this, there is no way to know exactly how many correct answers you need in order to pass.
- **Time management.** During the exam, you will have **2 hours** to answer **100 questions**, which means you have about **72 seconds on average per question**. If you are stuck between two very close answers, instead of spending significant extra time trying to decide, mark the question and come back to it later. Often you will find it much easier to answer such questions when you return to them with a fresh perspective.
- **The test itself may help you.** If you forget something and struggle to answer a question, note that questions or answer choices you see later in the exam can help you remember the right answer to a previous question.
- **Answer every single question.** Leaving a question blank will not award any points, whereas making your best guess may result in the correct answer.
- **Test early in the day.** Schedule your exam for the morning and get a good night's sleep the night before. The exam requires strong concentration, and taking it with a fresh mind is the easiest way to boost your score. Don't study on exam day — take the test with a clear and refreshed mind.
- **Be prepared for distractions.** If you are taking your exam at a testing center, be prepared for potential noise distractions. Even though most settings will be quiet, some may not be. These centers offer a variety of exams, so expect to see people entering and exiting the testing area throughout your exam.
- **Manage test anxiety.** If you feel anxiety during the exam, pause for a moment. Take a deep breath, hold it for five seconds, and then slowly release it. Repeat this process a few times if needed. Thinking about a very funny moment in your life also works well to de-stress before moving on. Without managing your anxiety, your mind will struggle to function effectively, making every question seem more difficult than it really is. If a challenging question heightens your anxiety, use this breathing exercise, flag the question for later review, and move on. You will often find that when you return to that question later, it will be much easier to answer.
- **Use all your time.** If you have extra time at the end of the exam, use it to review your answers. Start with the questions you flagged and those you feel most unsure about.

LEED FUNDAMENTALS

USGBC

Founded in the early 1990s and based in Washington, DC, the U.S. Green Building Council (USGBC) is a nonprofit organization committed to a sustainable future through cost- and energy-efficient green buildings. USGBC promotes buildings that are environmentally responsible, profitable, and supportive of human health.

USGBC's Mission:

"To transform the way buildings and communities are designed, built, and operated, enabling an environmentally and socially responsible, healthy, and prosperous environment that improves the quality of life."

USGBC's Vision:

"Buildings and communities will regenerate and sustain the health and vitality of all life within a generation."

USGBC accomplishes this mission through the following:

- LEED: The most widely known and used green building program across the world.
- Credentials: The LEED Green Associate credential and the various LEED AP credentials offer professionals a designation to help them stand out in the building industry.
- Greenbuild International Conference & Expo: The world's largest expo, launched in 2002, dedicated to green buildings.
- Education: USGBC provides educational programs about green buildings to further spread environmental principles and design.
- Advocacy: At every level of government, USGBC provides policymakers and community leaders with all the tools, strategies, and resources necessary to inspire action toward a sustainable built environment.
- Chapters & Committees: USGBC maintains a network of regional chapters and community organizations across the country for involvement from the local level to the national level.

GBCI

Green Business Certification Inc. (**GBCI**), formerly known as Green Building Certification Institute (GBCI), changed its name on April 16, 2015; is the sister organization of USGBC that administers LEED building certifications and LEED professional accreditations.

When a project submits all the necessary documents for a LEED application, GBCI is the organization that will review the application and determine if all the requirements set forth in a LEED rating system are fully met or not. And if the requirements are fulfilled, GBCI will grant the LEED certificate.

LEED professional exams are also administered by GBCI, and LEED professional credentials are awarded according to the exam results.

LEED FOR BUILDING DESIGN AND CONSTRUCTION (LEED BD+C) V5 RATING SYSTEMS

All types of buildings—whether newly constructed or existing buildings undergoing major renovations—will fall under the LEED BD+C category. Major renovations are defined as buildings with extensive interior alteration work in addition to work on the exterior shell and/or primary structural components and/or the core and peripheral mechanical, electrical, plumbing, and service systems and/or site work.

1) LEED BD+C: New Construction and Major Renovations

This is the most widely used rating system under LEED BD+C. It addresses design and construction activities for both new buildings and major renovations of existing buildings. To be eligible, **at least 60% of the project's gross floor area must be complete** by the time of LEED certification. **Complete spaces** are those that have their basic floor, wall, and ceiling finishes installed along with essential mechanical, electrical, and plumbing systems or fixtures necessary to occupy the space for its intended use.

2) LEED BD+C: Core and Shell Development

This is for projects in which the developer controls the design and construction of the building core and the entire mechanical, electrical, plumbing, and fire protection systems, but does not complete the interior tenant fit-out. To be eligible for this rating system, **at least 40% of the project's gross floor area must be incomplete** at the time of certification. **Incomplete spaces** are those that do not have their basic floor, wall, and/or ceiling finishes installed or essential mechanical, electrical, or plumbing systems or fixtures necessary to occupy the space for its intended use.

MINIMUM PROGRAM REQUIREMENTS (MPR)

The **minimum program requirements** are the minimum standards/characteristics that every LEED project should meet to pursue LEED certification. MPRs state whether a project can pursue LEED certification or not. There are three MPRs:

1) Every project must be in a permanent location on existing land

A project that is designed to move at any point in its lifetime is not eligible for LEED certification. Prefabricated or modular structures can be LEED-certified as long as they are installed permanently.

2) Must use reasonable LEED boundaries

The LEED project boundary must include all contiguous land that is associated with the project and that supports its typical operations. These will include the landscaping, septic or stormwater treatment equipment, parking, sidewalks, and even more. Any project space cannot be shown to be excluded in order to give the project an advantage in complying with credit requirements.

3) Must comply with project size requirements

All LEED projects must meet the following size requirements according to their rating system:

- LEED BD+C and LEED O+M rating systems: The project must include a **minimum of 1,000 square feet (93 square meters) of gross floor area.**
- LEED ID+C rating systems: The project must include a **minimum of 250 square feet (22 square meters) of gross floor area.**

LEED IMPACT AREAS AND CREDIT POINT ALLOCATION PROCESS

LEED v5 is structured around **three impact areas** which are the overarching objectives that every LEED project aims to accomplish, and every prerequisite and credit in the rating system is connected to one or more of them. The three impact areas are:

1) Decarbonization (50% of available points): This is the dominant impact area in LEED v5, reflecting the urgency of the climate crisis and the building industry's significant contribution to greenhouse gas emissions. Decarbonization targets all major sources of carbon emissions from buildings: operational carbon from energy use, embodied carbon from materials and construction, refrigerant emissions from mechanical systems, and transportation emissions from occupant commuting.

2) Quality of Life (25% of available points): This impact area recognizes that buildings are fundamentally about people. It addresses the health, well-being, resilience, and equity of building occupants and surrounding communities. LEED v5 requires project teams to understand who uses the building, assess the needs of the local community, evaluate natural hazards that could affect the site, and consider the workers and supply chains behind building materials.

3) Ecological Conservation and Restoration (25% of available points): This impact area focuses on limiting the environmental degradation caused by building construction and operations and promoting the rehabilitation and restoration of ecosystems. LEED v5 takes a comprehensive approach to ecological health by incorporating strategies that conserve natural habitats, minimize impacts on land and wildlife, restore ecosystems to enhance resilience and biodiversity, and manage natural resources responsibly.

Each credit in LEED makes a contribution to one or more of these impact areas. Credits that most directly address the prioritized impacts are given the greatest weight—meaning they contain more points.

LEED CERTIFICATION PROCESS FOR LEED BD+C PROJECTS

1. PROJECT REGISTRATION

The first step for LEED-certifying a building is to register the project through Arc, which means completing the registration form and paying the **registration fee** that varies by membership status. In this phase the project team should select the appropriate LEED rating system.

The certification platform defines two key roles: the owner and the LEED Project Administrator. The **owner** of the project refers to the individual(s) or entity(ies) that has the legal authority to fulfill the obligations defined in the LEED Certification Agreement. A **LEED Project Administrator** is the primary project contact for GBCI. The LEED Project Administrator is the team member who acts as a project manager, overseeing the LEED project and organizing team members for certain tasks, credits, or prerequisites.

2. APPLICATION

After the project registration phase, the project team will identify LEED credits to pursue and assign responsibility for the various credits to project team members. The team will then collect information, perform calculations, prepare documentation, fill out the credit forms demonstrating the achievement of the prerequisites and credits, and upload all the documentation to Arc. Once everything is complete, the LEED Project Administrator will verify the submission and the project team will pay the review (certification) fees to submit the application.

3. REVIEW

There are two types of review options for LEED BD+C projects: combined review and split review.

Combined review:

At the end of the construction phase:

1. Both design and construction submittals are sent by the project team.
2. There's a **preliminary review** by GBCI.
3. If additional information is needed to evaluate the application, GBCI will ask for clarifications and the project team will send the clarifications.
4. GBCI will proceed with the **final review** to evaluate the clarifications and to report the awarded/denied prerequisites and credits to the project team.
5. The project team will either accept or appeal the result.

Split review:

During the design phase:

1. Design submittals are sent by the project team.
2. **Preliminary design review** is conducted by GBCI.
3. If additional information is needed to evaluate the application, GBCI will ask for clarification, and the project team will send the clarification.
4. GBCI will proceed with the **final design review** to evaluate the clarifications and report the anticipated/denied prerequisites/credits to the project team.
5. The project team will either accept or appeal the result.

At the end of the construction phase the following occurs:

6. Both construction and extra design submittals (if applicable) are sent by the project team.
7. A **preliminary construction review** is conducted by GBCI.
8. If additional information is needed to evaluate the application, GBCI will ask for clarification, and the project teams will send the clarification.
9. GBCI will proceed with the **final construction review** to evaluate the clarifications and report the awarded/denied prerequisites/credits to the project team.
10. The project team will either accept or appeal the result.

4. CERTIFICATION OR DENIAL

This is the last step of the LEED certification. GBCI issues the final certification report, which will show the level of LEED certification awarded unless the certification is denied. The project teams can again appeal the results if needed, and if no appeals are made, then the project will be deemed "**closed out,**" which means the project team will no longer be able to appeal the certification level or review decisions for specific credits or prerequisites.

The total number of earned points will determine the level of LEED certification, assuming all the prerequisites and MPRs are met.

- LEED Certified™: 40–49 points
- LEED Silver: 50–59 points
- LEED Gold: 60–79 points
- LEED Platinum: 80–80+ points

LEED v5 Platinum projects must also meet specific decarbonization requirements in designated credits. For LEED BD+C: New Construction projects, for example, Platinum certification requires meeting minimum thresholds in four specific credits:

- Electrification Credit — Five points are required.
- Enhanced Energy Efficiency Credit — Eight points are required.
- Renewable Energy Credit — 100% of site energy use from any combination of Tier 1, Tier 2, and Tier 3 renewable energy.
- Reduce Embodied Carbon Credit — 20% reduction in embodied carbon.

Similarly, LEED BD+C: Core and Shell projects pursuing Platinum certification must also satisfy minimum performance thresholds in the same decarbonization-focused credits:

- Electrification Credit — Four points are required.
- Enhanced Energy Efficiency Credit — Five points are required.
- Renewable Energy Credit — 100% of base building energy use from any combination of Tier 1, Tier 2, and Tier 3 renewable energy.
- Reduce Embodied Carbon Credit — 20% reduction in embodied carbon.

PREREQUISITE & CREDIT SUMMARIES

CLIMATE RESILIENCE ASSESSMENT—PREREQUISITE

Complete a climate and natural hazard assessment. Identify observed, projected, and future natural hazards that could affect the site and building function (e.g., drought, extreme heat/cold, flooding, hurricanes and high winds, hail, landslides, sea level rise and storm surge, tornadoes, tsunamis, wildfires and smoke, winter storms). Identify **at least two priority** hazards to address through design. For each priority hazard, assess and specify:

- Intergovernmental Panel on Climate Change (IPCC) emissions scenario used (specifying the shared socioeconomic pathways)
- Projected service life of the project (e.g., FY2050 or 100 years)
- Hazard level; hazard risk rating
- Exposure, sensitivity, adaptive capacity, vulnerability, and overall risk levels
- Potential impact on site and building function, and on the site during construction

Where possible, use the assessment to inform planning, design, and O&M.

HUMAN IMPACT ASSESSMENT—PREREQUISITE

Complete a human impact assessment drawing on relevant information from the following four categories, as applicable:

- **Demographics**: race/ethnicity, gender, age, income, employment rate, population density, education, household types, nearby vulnerable populations.
- **Local infrastructure and land use**: adjacent transportation/pedestrian infrastructure, adjacent diverse uses, local/regional sustainability goals, accessibility codes.
- **Human use and health impacts**: housing affordability/availability, access to social services, community safety and groups, supply chain and construction workforce protections.
- **Occupant experience**: daylight, views, operable windows; air and water conditions; adjacent soundscapes, lighting, and wind patterns.
- Other: (specify).

Where possible, use the assessment to inform planning, design, and O&M.

CARBON ASSESSMENT—PREREQUISITE

USGBC provides the project team with a 25-year projection of the project's emissions from operations, refrigerants, and embodied carbon. The assessment uses data from the Operational Carbon Projection and Decarbonization Plan prerequisite, the Fundamental Refrigerant Management prerequisite, the Quantify and Assess Embodied Carbon prerequisite, and (optionally) the Transportation Demand Management credit.

TENANT GUIDELINES—PREREQUISITE

Create tenant guidelines to be shared with all tenants before signing the lease, including:

- A description of the sustainable design and construction features in the Core and Shell project plus the project's sustainability goals/objectives, including those for tenant spaces.
- Guidance and recommendations for incorporating sustainable strategies, products, materials, and services in tenant spaces (refer to the base building's attempted LEED prerequisites and credits for content requirements).
- A point of contact, from the base building team or new owner, for further coordination of base building design and construction documentation.

INTEGRATIVE DESIGN PROCESS—CREDIT (1 POINT)

Beginning in predesign and continuing through early occupancy, identify and apply opportunities for synergies across disciplines and building systems via:

- **Integrated team**: Assemble and convene an interdisciplinary team with diverse perspectives; ensure an equitable team effort through organized facilitation.
- **Design charette**: During predesign or early design, conduct a charette with the owner (or representative) and participants representing **at least four key perspectives** (e.g., architect, contractor, energy modeler, community engagement representative). To run an effective charrette, project teams can refer to the **U.S. Department of Energy's Handbook for Planning and Conducting Charrettes for High-Performance Projects**, which provides checklists and sample agendas.
- **LEED goal setting**: Define specific, measurable project goals addressing the LEED v5 impact areas (decarbonization, quality of life, ecological conservation and restoration) and incorporate them into the owner's project requirements.

GREEN LEASES—CREDIT (1–6 POINTS)

OPTION 1: STANDARD GREEN LEASE (1–6 POINTS)

Develop a standard green lease setting tenant fit-out standards and commit to including it in each future lease.

Tenant requirements must include standards ensuring compliance, for any tenant-installed systems, with the Climate Resilience Assessment prerequisite, the Human Impact Assessment prerequisite, the Carbon Assessment prerequisite, the Minimum Water Efficiency prerequisite, the Minimum Energy Efficiency prerequisite, the Fundamental Commissioning prerequisite, the Energy Metering and Reporting prerequisite, the Planning for Zero Waste Operations prerequisite, and the Fundamental Air Quality prerequisite.

AND/OR

OPTION 2: EXECUTED STANDARD GREEN LEASE (1 POINT)

Provide documentation of an executed tenant green lease or tenant letter of attestation meeting Option 1's criteria.

AND/OR

OPTION 3: GREEN LEASE LEADERS RECOGNITION (1–6 POINTS)

Achieve Green Lease Leaders recognition at the Silver, Gold, or Platinum recognition level. The Green Lease Leaders program is an initiative developed by the **Institute for Market Transformation** (**IMT**) with support from the **U.S. Department of Energy**. The program is open to both U.S. and international companies, and its recognition criteria align with the EPA's **ENERGY STAR Tenant Space recognition program**.

SENSITIVE LAND PROTECTION—CREDIT (1 POINT)

OPTION 1: PREVIOUSLY DEVELOPED SITES (1 POINT)

Locate the entire development footprint on land that has been previously developed.

OR

OPTION 2: PREVIOUSLY UNDEVELOPED SITES (1 POINT)

Locate the development footprint on land that does NOT meet any of the following criteria for sensitive land:

- **Prime farmland** (as defined by the **U.S. Department of Agriculture** in the U.S. Code of Federal Regulations and identified in **NRCS soil survey**).
- **Floodplains** — Areas shown as floodplains on **Federal Emergency Management Agency** (**FEMA**) flood hazard maps or maps from local authorities. If no maps are available, the project team must confirm the site is not in an area with a 1% or greater annual chance of flooding in any year (also known as **100-year floodplain**).
- **Notable habitat** — for threatened/endangered species under the **U.S. Endangered Species Act**, or NatureServe GH/G1/G2 species/communities.
- **Water bodies** — on or within **100 feet** (30 meters), except minor improvements.
- **Wetlands** — on or within **50 feet** (15 meters), except minor improvements.
- **Steep slopes** — protect 40% of slopes between 15%–25%; protect 60% of slopes steeper than 25% from all development.

EQUITABLE DEVELOPMENT—CREDIT (1–2 POINTS)

OPTION 1: PRIORITY SITES (1–2 POINTS)

PATH 1: BROWNFIELD REMEDIATION

Locate the project on a brownfield where soil or groundwater contamination has been identified. For voluntary remediation by the project team, obtain confirmation from the authority verifying that the site is a brownfield, and perform remediation to the authority's satisfaction.

OR

PATH 2: HISTORIC LOCATION

Locate the project in a historic district identified by the local government based on a growth management plan or policy.

AND/OR

OPTION 2: HOUSING AND JOBS PROXIMITY (1–2 POINTS)

PATH 1: SUPPORT LOCAL ECONOMY

Employ individuals who live within the administrative district of the project site for **15%** of the construction jobs created by the project.

OR

PATH 2: LOCATION-EFFICIENT AFFORDABLE HOUSING

This path focuses on affordable housing and is available for residential or mixed-use projects. It requires that a portion of new dwelling units be provided as affordable rentals and/or for-sale homes, priced for households earning below the **area median income** (**AMI**). Rentals must stay affordable for at least 15 years.

At least 10% of the residential floor area must be affordable, with rental units priced **up to 60% AMI** or for-sale units priced up to 80% AMI. Additionally, the project team should meet Option 2 (Access to Transit) or Option 3 (Walkable Location) of the Compact and Connected Development credit, or be located in a community with a jobs-to-housing ratio above 1:2 within a half-mile walk.

OR

OPTION 3: EQUITABLE CONSTRUCTION (2 POINTS)

Provide access to workforce development training for construction workers through one of the following:

- Job-related skills training: on-the-job training in a Department of Labor registered apprenticeship program demonstrating that **15% or more** of total project construction hours were performed by participants enrolled in registered apprenticeship programs.
- Life-skills training: programs for construction workers conducted by an organization or government entity on the construction site demonstrating scheduling of one course per month for the duration of construction.

OR

OPTION 4: EQUITABLE ACCESS TO RESOURCES — SCHOOLS ONLY (2 POINTS)

Schools may meet Options 1, 2, and/or 3, or instead pursue this option.

PATH 1: PUBLIC USE SPACES

Ensure that **at least three** of the following types of spaces in the school are accessible to and available for shared use by the public: auditorium, gymnasium, cafeteria, one or more classrooms, playing fields and stadiums, or joint parking. Provide access to toilets in joint-use areas after normal school hours.

AND/OR

PATH 2: COMMUNITY PARTNERSHIP

Contract with the community or other organizations to provide **at least two types** of dedicated-use spaces in the building, such as a commercial office, health clinic, community service centers provided by state or local offices, library or media center, parking lot, or one or more commercial businesses. Provide access to toilets in joint-use areas after normal school hours.

AND

OPTION 5: SENSITIVE PROJECT LOCATION — DATA CENTERS, WAREHOUSES, AND DISTRIBUTION CENTERS ONLY (1–2 POINTS)

Data Centers, Warehouses, and Distribution Centers must meet Options 1, 2, and/or 3 and also satisfy this option.

Locate the project building a minimum of **300 feet** (90 meters) away from the property lines of the nearest sensitive receptors, such as residential areas, schools, and public parks.

COMPACT AND CONNECTED DEVELOPMENT—CREDIT (1–6 POINTS)

OPTION 1: SURROUNDING DENSITY (1–2 POINTS)

Under this option, the project must be located on a site where the existing surrounding density within a 0.25-mile (400-meter) offset of the project boundary meets LEED's thresholds. Projects in areas with higher surrounding density can earn more points.

AND/OR

OPTION 2: ACCESS TO TRANSIT (1–4 POINTS)

PATH 1: PUBLIC TRANSIT SERVICE

Locate any functional entry of the project within **0.25 miles** (400 meters) walking distance of existing or planned bus, streetcar, or informal transit stops, OR within **0.5 miles** (800 meters) walking distance of existing bus rapid transit stops, passenger rail stations (light, heavy, or commuter rail), or commuter ferry terminals.

OR

PATH 2: PROJECT-SPONSORED TRANSIT SERVICE

Commit to year-round transit service (such as vans, shuttles, or buses) for regular occupants and visitors between the project site and external destinations like residential areas and public transportation stations, guaranteed for **at least three years** from the project's certificate of occupancy. Provide at least one accessible transit stop shelter within **0.25 miles** (400 meters) walking distance of a functional entry.

AND/OR

OPTION 3: WALKABLE LOCATION (1–3 POINTS)

Locate on a site that meets the location-efficiency thresholds below via Walk Score or by proximity to existing, publicly available uses within **0.5 miles** (800 meters) walking distance of any functional entry.

For Schools, teams can meet Options 1, 2, and/or 3, and may also pursue Options 4 and/or 5.

OPTION 4: SURROUNDING DENSITY AND DEVELOPMENT

PATH 1: SURROUNDING DENSITY

Meet Option 1: Surrounding Density requirements described for New Construction.

AND/OR

PATH 2: CONNECTED SITE

Locate the project on a previously developed site that also meets one of the connected-site conditions below.

- Adjacent site: at least a contiguous **25%** of the project boundary borders parcels that are previously developed sites.
- Infill site: **at least 75%** of the project boundary borders parcels that are previously developed sites.

Bordering rights-of-way do not constitute previously developed land — it is the status of the property on the other side of the right-of-way that contributes to the calculation — and any part of the boundary that borders a water body is excluded.

AND/OR

OPTION 5: ACCESS TO TRANSIT OR PEDESTRIAN ACCESS

PATH 1: ACCESS TO TRANSIT

Meet Option 2: Access to Transit described above.

OR

PATH 2: PEDESTRIAN ACCESS

Locate the project within an attendance boundary where dwelling units are within **0.75 miles (1,200 meters)** walking distance of a functional entry of a school building for students in eighth grade or below or ages 14 and below, and within **1.5 miles (2,400 meters)** walking distance of a functional entry of a school building for students in ninth grade or above or ages 15 and above. Provide pedestrian access to the site from all residential areas in the attendance boundary.

AND/OR

OPTION 6: SURROUNDING DEVELOPMENT AND RESOURCES — DATA CENTERS, WAREHOUSES, AND DISTRIBUTION CENTERS ONLY (1–2 POINTS)

For data centers, warehouses, and distribution centers, teams can meet Options 1, 2, and/or 3, and may also pursue Option 6.

PATH 1: DEVELOPMENT AND ADJACENCY

Locate the project on a site that meets one of the conditions below.

- Previously developed site that was used for industrial or commercial purposes.
- Previously developed and adjacent site with bordering parcels currently used for industrial or commercial purposes.

To qualify as an adjacent site, at least a contiguous 25% of the project boundary must border parcels that are previously developed sites. Bordering rights-of-way do not constitute previously developed land, and any part of the boundary bordering a water body is excluded.

AND/OR

PATH 2: TRANSPORTATION RESOURCES

Locate the project on a site that has the transportation resources below. Provide at least two of the four resources; providing all four earns the higher amount within this option.

- The site is within a 10-mile (16-kilometer) driving distance of a main logistics hub.
- The site is within a one-mile (1,600-meter) driving distance of an on-off ramp to a highway.
- The site is within a one-mile (1,600-meter) driving distance of an access point to an active freight rail line.
- The site is served by an active freight rail spur.

A planned resource must be sited, funded, and under construction by the date of the certificate of occupancy and complete within 24 months of that date.

AND/OR

OPTION 7: SURROUNDING DENSITY — HEALTHCARE ONLY (1 POINT)

Healthcare projects may meet Options 1, 2, and/or 3 and may also pursue this option. Locate on a site where the surrounding existing density within a 0.25-mile (400-meter) offset of the project boundary meets one of the following: at least 7 dwelling units per acre (17.5 dwelling units per hectare) with a 0.5 floor-area ratio (the counted density must be existing density, not zoned density), OR at least 22,000 square feet per acre (5,050 square meters per hectare) of buildable land.

For previously developed existing rural healthcare campus sites, achieve a minimum development density of 30,000 square feet per acre (6,890 square meters per hectare).

TRANSPORTATION DEMAND MANAGEMENT—CREDIT (1–4 POINTS)

Complete the transportation demand assessment, then implement parking and active-travel measures to earn points. Assess the vehicle miles traveled and carbon emissions associated with regular building occupants' travel to and from the project by estimating annual VMT, estimate an annual baseline case for carbon emissions, assess low-carbon transportation options, estimate an annual proposed case for carbon emissions, and estimate the total reduction between the baseline and proposed cases.

Projects that participate in a local or regional government-mandated TDM program satisfy this assessment. Residential affordable housing projects in an infill location, and office, mixed-use, residential, or retail projects located within a transit priority area or within 0.5 miles (800 meters) walking distance of an existing or planned major transit stop, are exempt from the assessment requirements.

After the assessment, implement one or more of the following.

OPTION 1: PARKING (1–3 POINTS)

PATH 1: REDUCE PARKING

Reduce parking capacity relative to the base ratios in the **Institute of Transportation Engineers Parking Generation Manual**, sixth edition (or a comparable resource applied by a qualified transportation engineer or planner, or supplementary LEED guidance). Achieve **at least a 30% reduction** from base ratios; greater reductions earn more within this option, with a 60% reduction and a 100% reduction (no parking) reaching the higher tiers.

AND/OR

PATH 2: PARKING FEE

Implement a daily, monthly, or annual parking fee at a cost equal to or greater than the local market rate for public or private parking.

AND/OR

OPTION 2: ACTIVE TRAVEL FACILITIES (1–3 POINTS)

PATH 1: BICYCLE NETWORK AND STORAGE

- Bicycle network: design or locate the project such that a functional entry and/or bicycle storage is within a **600-foot** (180-meter) walking or bicycling distance of a bicycle network that provides a contiguous network spanning **at least three miles** (4,800 meters) and consists of bicycle paths, lanes, multiuse trails, or streets with a maximum speed limit of 25 mph (40 kph). Sidewalks where local code permits bicycles are acceptable. Planned bicycle trails or lanes may be counted if fully funded by the date of the certificate of occupancy and scheduled for completion within **three years** of that date. For schools, provide dedicated bicycle lanes that extend from the student bike-parking location to at least the end of the school property without any barriers.
- Short-term bicycle storage: provide short-term storage within **600 feet** (180 meters) walking distance of any main entrance, but **no fewer than four storage spaces** per building.

- Long-term bicycle storage: provide long-term storage within **300 feet** (90 meters) walking distance of any functional entry, but **no fewer than four storage spaces** per building, in addition to the short-term storage spaces.
- Shared micromobility storage, bicycle-sharing stations, and/or publicly available bicycle parking may count for up to 50% of the required short-term and long-term storage if it meets the maximum allowable walking distance, is not double counted, and its location is communicated to occupants and visitors.

AND/OR

PATH 2: SHOWER AND CHANGING FACILITIES

Under this path, the project should include showers and changing facilities for **1% of regular building occupants**. Off-site facilities may be used if they serve all occupants and are within **0.25 miles (400 meters)** walking distance.

AND/OR

PATH 3: BICYCLE MAINTENANCE

Provide a permanent and secure bicycle repair station that includes a complete set of tools and an air pump securely fastened to the repair stand in the area dedicated to long-term bicycle storage.

ELECTRIC VEHICLES—CREDIT (1–2 POINTS)

OPTION 1: ELECTRIC VEHICLE SUPPLY EQUIPMENT (1–2 POINTS)

- Install electric vehicle supply equipment (EVSE) that provides **Level 2** or **Level 3** charging capacity per the manufacturer's requirements and the National Electrical Code (NFPA 70); provides 208–240 volts or greater for each required space; meets the connected functionality criteria for **ENERGY STAR-certified EVSE** and is capable of responding to time-of-use market signals such as price; and includes at least one EV charging station with an accessible parking space at least 9 feet (2.5 meters) wide with a 5-foot (1.5-meter) access aisle and accessibility features for people with mobility, ambulatory, and visual limitations.
- Commercial parking: equip **at least 5%** of total parking spaces or **at least two spaces** (whichever is greater); equipping 10% or at least four spaces (whichever is greater) earns more within this option.
- Residential parking: equip **at least 10%** of total parking spaces or **at least five spaces** (whichever is greater); equipping 15% or at least 10 spaces (whichever is greater) earns more within this option.

AND/OR

OPTION 2: ELECTRIC VEHICLE READINESS (1 POINT)

- Make the minimum number of total parking spaces EV-ready, providing a full-circuit installation including 208–240 volts, a 40-amp panel capacity, and a conduit (raceway) with wiring that terminates in a junction box or charging outlet. Any space with installed EVSE counted under the Electric Vehicle Supply Equipment option may not also be counted as an EV-ready space here.

- Commercial parking: **at least 10%** of total parking spaces or **at least 10 spaces**, whichever is greater.
- Residential parking: **at least 20%** of total parking spaces or **at least 20 spaces**, whichever is greater.

MINIMIZED SITE DISTURBANCE—PREREQUISITE

This prerequisite requires the project to limit construction-related disturbance to the site. Two components must both be met: an erosion and sedimentation control plan and a preconstruction site assessment.

1) Erosion and Sedimentation Control Plan:

- Conformance — Create and implement an erosion and sedimentation control plan for all construction activities that conforms to the erosion and sedimentation requirements of the **U.S. EPA's 2022 Construction General Permit (CGP)**; EU Taxonomy: DNSH, Pollution Prevention, Item 4 Noise and Dust; or the local equivalent. Projects must apply the CGP regardless of size.
- Construction-exclusion zones — Establish construction-exclusion zones demarcated by physical barriers and stormwater controls to protect any identified critical habitat for threatened or endangered species from discharges and discharge-related activities.
- Inspections — Inspect all controls and management practices at least every 7 days, or every 14 days plus within 24 hours of a storm that drops 0.25 inches (6 millimeters) or more of rain, and must inspect dewatering once on each day it discharges.
- Corrective action — Take immediate corrective action to repair or replace controls when they are failing.

AND

2) Site Assessment:

Collect information about the site in a preconstruction survey or assessment that informs the site design and demonstrates how the site features influenced the project design, addressing the following as applicable:

- Special-status vegetation — Conserve **100%** of special-status vegetation located on-site, as defined by local, state, or federal entities.
- Healthy habitat — Identify healthy plant communities and implement strategies to minimize damage to these areas during construction and ongoing activities, establishing exclusion zones demarcated by physical barriers to minimize intrusion or disturbance.
- Invasive vegetation — Map existing invasive species on-site and manage their removal and control before and during construction. Teams may plant only native or adapted vegetation that is not classified as invasive and is suited to the project's **EPA Level III ecoregion**.

BIODIVERSE HABITAT—CREDIT (1–2 POINTS)

OPTION 1: PRESERVE AND RESTORE HABITAT (1–2 POINTS)

PATH 1: GREENFIELD SITES

- Preserve **40%** of the greenfield area on the site by protecting these areas from all development and construction activity.

OR

PATH 2: PREVIOUSLY DISTURBED SITES

- Meet the requirements of Path 1 (Greenfield Sites) if such areas exist, and restore previously disturbed areas of the site. Restore **at least 20%** of the previously disturbed area (10% for zero lot line projects); restoring 40% (20% for zero lot line projects) earns the higher amount within this option. Dedicated athletic fields used solely for athletic use are exempt from the total site area and may not count toward protected greenfield or restored habitat areas.
- Soil restoration — Restore all on-site soils disturbed by previous development and by current construction that will serve as final vegetated areas. Any imported soils must be reused comparably to their original function and may not include soils defined by the **NRCS web soil survey** (or local equivalent) as prime farmland, unique farmland, or farmland of statewide or local importance; soils from other greenfield sites; or sphagnum peat moss or organic amendments containing sphagnum peat. Engineered growing medium for vegetated roofs is exempt.
- Vegetation restoration — Plant native and adapted vegetation that is not currently listed as invasive, including **at least 10 species** native or adapted to the project's **EPA Level III ecoregion** (or local equivalent); a minimum of two of the categories trees, shrubs, and ground cover (zero lot line projects are exempt from this requirement); and at least 110 square feet (10 square meters) of native flowering plants appropriate for local pollinators, in groupings of at least 10 square feet (one square meter), with the pollinator habitat area designated by signage.

AND/OR

OPTION 2: BIRD-FRIENDLY GLASS (1 POINT)

- Glass used below specified heights on the building and site exterior must have a maximum **threat factor of 30**, as defined in the **American Bird Conservancy's Threat Factor Database**. This applies to all glass, including spandrel glass, from grade up to 50 feet (15 meters) measured at all points; up to 20 feet (six meters) measured from the finished grade of a green roof; and at any distance from grade or roof for glass in guardrails and windshields.

ACCESSIBLE OUTDOOR SPACE—CREDIT (1 POINT)

Meet the following three components.

1) Sufficient Outdoor Space Area

Provide barrier-free and physically accessible outdoor space for people with limited mobility that is greater than or equal to **30% of the total site area** (including building footprint). **At least 25%** of the required outdoor space must be vegetated and planted with two or more types of vegetation or have an overhead vegetative canopy.

AND

2) Urban Outdoor Space

Include one or more of the following elements:

- Biophilic space — An area that meets the vegetation restoration requirements of the Biodiverse Habitat credit and includes elements of human interaction, such as observation platforms or paths.
- Garden — Space dedicated to community gardens or urban food production.
- Recreational area — Recreation-oriented paving or landscape area that encourages physical activity, such as courts, fields, a track, play space, or swimming pools.
- Social area — Pedestrian-oriented paving or landscape area that accommodates outdoor social activities and includes seating for **5%** of occupants.

AND

3) Community Outdoor Space

Include one or more of the following elements:

- Community — Publicly accessible during daylight hours and open to all members of the community.
- Cultural — Include at least two art installations or sculptures by local artists.
- Acoustics — Include elements that provide positive soundscapes if located within **0.25 miles (400 meters)** of a significant noise source, such as a roadway, airport, or rail line.

RAINWATER MANAGEMENT—CREDIT (1–3 POINTS)

OPTION 1: PERCENTILE OF RAINFALL EVENTS (1–3 POINTS)

Retain on-site the runoff from the associated percentile of regional or local rainfall events in a manner that best replicates natural site hydrology, using low-impact development (LID) and green infrastructure (GI) practices that may be structural or nonstructural. Retain at least the 80th percentile event (70th percentile for zero lot line projects); retaining the 85th percentile (75th for zero lot line) or the 90th percentile (80th for zero lot line) earns more within this option.

Water reuse bonus — Projects that collect and reuse a portion of the chosen percentile event volume to meet one or more building and grounds end uses earn an additional amount within this option for the lower retention tiers (the highest retention tier has no additional reuse bonus). Eligible end uses include irrigation, flush fixtures, makeup water systems such as cooling towers or boilers, and other process water demands.

OR

OPTION 2: NATURAL LAND COVER CONDITIONS (3 POINTS)

Calculate the difference between the projected runoff volume under the proposed design conditions and the runoff volume under the natural land cover conditions that existed prior to any disturbance, and retain on-site the increase in runoff volume using LID and GI practices.

Zero lot line projects only — Project teams may combine on-site and off-site strategies to retain runoff from the associated percentile regional or local rainfall event, engaging local or regional authorities to coordinate off-site strategies such as community-wide rainwater management programs.

ENHANCED RESILIENT SITE DESIGN—CREDIT (2 POINTS)

Project teams should design and construct the site and site structures to meet the following best practices for **at least two** of the highest-priority hazards identified under Climate Resilience Assessment prerequisite.

1) Drought:

Comply with the Enhanced Water Efficiency credit requirements, specify native and/or drought-tolerant adapted plantings, and ensure water makeup for any created water features complies with SITES C3.4 or local equivalent. For Core and Shell projects, water makeup for created water features must not exceed 5,000 gallons (18,927 liters) of potable water per year, or 75% of annual makeup must come from alternative water sources.

AND/OR

2) Extreme heat:

Integrate two or more additional elements from the following list:

- Provide shaded external spaces adjacent to buildings for use during extreme heat events.
- Provide evaporative cooling solutions such as fountains, misters, or water features.
- Orient buildings and massing to self-shade in summer and extreme heat conditions.
- Provide outdoor cooling stations with emergency backup power.
- Demonstrate proximity to an emergency cooling station within 0.25 miles (400 meters).
- Use paving materials with an initial solar reflectance (SR) value of at least 0.33.
- Use an open-grid pavement system that is at least 50% unbound.

AND/OR

3) Flooding:

Integrate two or more of the following strategies, in accordance with ASCE 24 and FEMA 543 standards or local equivalent:

- Critical utilities — Locate critical utilities in new construction above the design flood elevation (DFE) plus recommended freeboard; in retrofits, place them inside protective, floodproofed enclosures; design potable water systems to resist flood damage, infiltration, and effluent discharge; elevate on-site wellheads above the surrounding landscape; design sewage systems to avoid infiltration and backup; design and anchor plumbing conduits, water supply lines, gas lines, and electric cables that extend below DFE to resist flooding; and design and anchor rainwater storage tanks to resist flood forces.
- Ensure that all structural materials, finish materials, and connectors used below DFE are flood resistant.
- Certify the project under a qualifying flood-resilient design standard.

AND/OR

4) Hail:

Design and construct the site and site structures according to FORTIFIED Commercial High Wind and Hail Specific Design Requirements, or local equivalent.

AND/OR

5) Hurricanes and high winds:

For projects in hurricane-prone areas, design and construct the site and site structures according to FORTIFIED Commercial Wind standards, or local equivalent. For projects in high-wind areas, comply with wind design measures per ASCE/SEI 7-10 in specified FEMA zones or local equivalent. Additionally, install backup power systems in hurricane-prone regions; install electrical connections with a transfer switch or docking station (storm switch) to support backup power for critical mechanical and electrical systems; and create windbreaks using landscape forms, vegetation, and other locally appropriate natural systems.

AND/OR

6) Sea level rise:

Design and construct the site to accommodate flooding based on sea level rise and storm surge projections for the design service life of the project, and meet two or more of the following:

- Incorporate elevated foundations to minimize projected flood damage to buildings.
- Use materials resistant to projected water damage for construction.
- Apply sealants and coatings to prevent projected water infiltration into structures.
- Install flood barriers to block projected floodwater from entering buildings.
- Design GI solutions to manage projected storm surge and stormwater runoff.
- Ensure backup power systems are in place to maintain critical functions during projected flooding events.
- Develop integrated drainage systems to manage projected excess flooding.
- Engage in community-level planning, partnerships, and/or design workshops to coordinate flood mitigation efforts that effectively and equitably address vulnerable populations.
- Retrofit existing structures to enhance their resilience to future flood risks.

AND/OR

7) Tsunamis:

Mitigate the impact of tsunamis through site planning strategies described in Designing for Tsunamis (U.S. National Tsunami Hazard Mitigation Program), or local equivalent. Additionally, incorporate the following elements from the TsunamiReady Guidelines, or local equivalent: install tsunami danger area and evacuation route signage, and install Public-Alert-certified NOAA Weather Radio receivers in critical facilities and public venues.

AND/OR

8) Wildfires:

Follow wildfire management practices for wildland–urban interface design, vegetation management, debris disposal, and fire safety for equipment referenced in the National Wildfire Coordinating Group Standards for Mitigation 2023, or local equivalent. Design and construct the site and site structures in compliance with the SITES v2 rating system credit 4.11 (reduce the risk of catastrophic wildfire), or local equivalent, and reduce fuel using the zone concept in Safer from the Start, Appendix E, or local equivalent.

AND/OR

9) Winter storms:

Meet two or more of the following:

- Provide adequate ingress/egress for vehicles and snow removal equipment.
- Provide a snow removal plan, including compatible road materials, areas for accumulated snow, and roof snow removal.
- Ensure safe walking surfaces to exterior parking areas by considering heated sidewalks with renewable energy sources.
- Specify native or adapted planting with a capacity for heavy snow loads.

HEAT ISLAND REDUCTION—CREDIT (1–2 POINTS)

OPTION 1: NONROOF AND ROOF (1 POINT)

Use any combination of nonroof, high-reflectance roof, and vegetated roof strategies to meet the following formula:

$$\frac{\text{Area of Non-roof Measures}}{0.5} + \frac{\text{Area of High-Reflectance Roof}}{0.75} + \frac{\text{Area of Vegetated Roof}}{0.75} \geq \text{Total Site Paving Area} + \text{Total Roof Area}$$

- Nonroof measures — Shade over pavement areas measured in plain view at noon using existing or new plants (assuming 10-year canopy width) or vegetated structures, in place at the time of the occupancy permit; structures covered by energy generation systems such as solar thermal collectors, photovoltaics, and wind turbines; architectural devices or structures (if a roof, an aged SR value of at least 0.28 per ANSI/CRRC S100; if not a roof or if aged SR is unavailable, an initial SR of at least 0.33 at installation); paving materials with an initial SR value of at least 0.33; and open-grid pavement systems that are at least 50% unbound.
- High-reflectance roof — Use roofing materials with an aged solar reflectance index (SRI) value at or above the required minimum (or initial SRI if aged is unavailable): low-sloped roofs at a slope of 2:12 or less require an initial SRI of 82 or an aged SRI of 64, and steep-sloped roofs above 2:12 require an initial SRI of 39 or an aged SRI of 32. Roof area consisting of functional, usable spaces such as helipads, recreation courts, and swimming pools may meet the nonroof measures instead; applicable roof area excludes area covered by mechanical equipment, solar energy panels, skylights, and other appurtenances.
- Vegetated roof — Install a vegetated roof using native or adapted plant species.

AND/OR

OPTION 2: PARKING UNDER COVER (1 POINT)

Place **100%** of parking spaces under cover. Any roof used to shade or cover parking must have an aged SRI of at least 32 (or an initial SRI of at least 39 at installation if aged information is unavailable), be a vegetated roof, or be covered by energy generation systems such as solar thermal collectors, photovoltaics, and wind turbines. Calculations must include all existing and new off-street parking spaces subsidized, leased, or owned by the project, including parking outside the project boundary used by the project; on-street parking in public rights-of-way is excluded.

AND/OR

OPTION 3: TREE EQUITY (1 POINT)

For U.S. projects, evaluate the American Forests Tree Equity score for the site location; for projects in areas ranked "**high priority**" and "**highest priority**," use the results to inform an increase in on-site canopy cover from the existing condition.

LIGHT POLLUTION REDUCTION—CREDIT (1 POINT)

Meet the following uplight, light trespass, and internally illuminated exterior signage requirements for exterior luminaires located inside the project boundary. To determine the lighting zone of the project, project teams should refer to the **Illuminating Engineering Society and Dark Sky International (IES/IDA) Model Lighting Ordinance (MLO) User Guide.**

Uplight:

Do not exceed the maximum uplight rating for each Model Lighting Ordinance (MLO) lighting zone, based on the specific light source installed: zones LZ0 and LZ1 allow U0, LZ2 allows U2, LZ3 allows U3, and LZ4 allows U4.

AND

Light Trespass:

Do not exceed the maximum luminaire backlight and glare ratings, based on the installed light source, as defined in **IES TM-15-11, Addendum A**, according to the mounting location and distance from the lighting boundary.

AND

Internally Illuminated Exterior Signage:

Do not exceed the LEED's maximum nighttime luminance level for internally illuminated signage.

The following exterior lighting is exempt from the uplight and light trespass requirements, provided it is controlled separately from nonexempt lighting:

- Specialized signal, directional, and marker lighting for transportation
- Lighting used solely for façade and landscape lighting in MLO lighting zones 3 and 4 that is automatically turned off from midnight until 6 a.m.
- Lighting for theatrical purposes
- Government-mandated roadway lighting
- Lighting for hospital emergency departments
- Lighting for the national flag in lighting zones 2, 3, or 4
- Internally illuminated exterior signage

WATER METERING AND REPORTING—PREREQUISITE

Install or use existing permanent water meters to monitor, record, and report total water consumption for potable and alternative water sources serving the building and associated grounds. Report whole-project use for each type of water source supplied, with the following provisions.

- Data access — The facility manager and/or tenant(s) must be able to access the meter data.
- Separate metering — Meter alternative water sources separately from municipally supplied potable water.
- Data sharing — Commit to sharing the resulting whole-project water usage data with USGBC at least annually. This commitment must carry forward for five years or until the building changes ownership or lessee.

The requirements may be applied to the project scope of work and may exclude future tenant utility services and submeters that will be installed in the tenant scope of work.

MINIMUM WATER EFFICIENCY—PREREQUISITE

Projects should satisfy the minimum fixture and fittings efficiency, minimum equipment water efficiency, and minimum outdoor water efficiency requirements as follows.

1) Minimum Fixture and Fittings Efficiency:

Projects located where standard supply pressure differs from the LEED baseline supply pressure may calculate the water consumption of flow fixtures and fittings at the local standard supply pressure. Meet either option below for all new and existing fixtures and fittings within the project boundary.

OPTION 1: PRESCRIPTIVE PATH—MAXIMUM FLUSH AND FLOW RATES

Do not exceed the maximum installed flush or flow rates specified by LEED for each fixture and fitting. The **WaterSense** label is available for toilets, urinals, private lavatory faucets, and showerheads for projects in the U.S. and Canada.

OR

OPTION 2: PERFORMANCE PATH—CALCULATED REDUCTION

Reduce aggregate water consumption by **20%** from the baseline rates.

AND

2) Minimum Equipment Water Efficiency:

Newly installed appliances, equipment, and processes within the project boundary must meet the LEED specifications. Existing appliances and equipment can be excluded. Project teams should use **ENERGY STAR** or performance equivalent clothes washers, dishwashers, and ice machines.

AND

3) Minimum Outdoor Water Efficiency:

OPTION 1: NO IRRIGATION

Show that the landscape does not require a permanent irrigation system beyond a maximum **two-year establishment period**.

OR

OPTION 2: EFFICIENT IRRIGATION

Reduce the project's irrigation water requirement by **at least 30%** from the calculated baseline for the site's annual theoretical irrigation requirement (TIR). Reductions must be achieved through plant species selection and irrigation system efficiency, as calculated by the TIR methodology outlined by the U.S. Environmental Protection Agency.

WATER METERING AND LEAK DETECTION—CREDIT (1 POINT)

OPTION 1: SUBMETERS (1 POINT)

Install permanent water meters for each applicable subsystem below.

- Indoor plumbing fixtures and fittings — meter systems serving **at least 80%** of indoor fixtures and fittings as described in the Minimum Water Efficiency prerequisite. Fixtures and fittings not addressed in the prerequisite, including janitor sinks, water coolers, and bottle fillers, may be included or excluded at the project team's discretion.
- Irrigation system.
- Each makeup water system — for example, the cold water inlet for domestic hot water, swimming pools, chilled water systems, and process water systems.
- Commercial kitchen — if the kitchen serves at least 100 meals per day of operation.
- Laundry — if the project includes commercial laundry equipment that processes at least 120,000 lb (57,606 kg) of laundry per year, or if the project includes a public laundry room.

The facility manager and/or tenant(s) must be able to access the submeter data in real time via local network, building management system, cloud service, app, or online database. All submeters must be capable of recording data at least hourly.

Healthcare projects — additionally install water meters in any five of the following: purified water systems (reverse-osmosis, deionized); filter backwash water; dietary department; laundry; laboratory; central sterile and processing department; physiotherapy, hydrotherapy, and treatment areas; surgical suite; closed-loop hydronic system makeup water; and cold water makeup for domestic hot water systems. If fewer than five of these subsystems are within the project scope, the project may instead submeter all applicable water subsystems.

Residential projects — install a permanent water meter for each residential dwelling unit that measures total potable water use for the unit. These meters need not be utility owned or utility grade.

OR

OPTION 2: LEAK DETECTION SENSORS (1 POINT)

Install permanent water flow meters or sensors for each applicable subsystem below.

- Project irrigation system — at the point of entry, if irrigation is included in the project scope.
- Flush fixtures — **at least 50%** of the project flush fixtures. Water sensors can be installed on each flush fixture or for a group of flush fixtures (such as one per restroom facility).
- Each makeup water system — for example, the cold water inlet for domestic hot water, swimming pools, chilled water systems, and process water systems.

The leak detection system must be able to identify a leak triggered by an abnormal flow rate above the normal range, or physically detect a water leak, and initiate an alarm upon detection. The facility manager and/or tenant(s) must be able to access the sensor data in real time via local network, BMS, cloud service, app, or online database. Develop an action plan addressing how the building manager and/or tenant(s) will access data in real time and address and remedy any detected leak.

ENHANCED WATER EFFICIENCY—CREDIT (1–8 POINTS)

Pursue Option 1 on its own, or pursue any combination of Options 2 through 6.

OPTION 1: WHOLE-PROJECT WATER USE (1–8 POINTS)

Develop a water use baseline and create a proposed use model. Points scale with the percent reduction in overall project water use from the baseline, beginning at **30%**, up to 65%. Using alternative water to achieve the reductions raises each tier by one point.

OR

OPTION 2: FIXTURE AND FITTINGS—CALCULATED REDUCTION (1–3 POINTS)

Further reduce fixture and fitting water use from the calculated baseline used in the performance path of the Minimum Water Efficiency prerequisite. Additional potable water savings above the prerequisite level can be earned using alternative water sources. Points scale with the percent reduction from baseline (30%, 35%, 40%).

AND/OR

OPTION 3: APPLIANCE AND PROCESS WATER (1–2 POINTS)

Newly installed equipment within the project boundary must meet the minimum requirements for commercial washing machines, commercial kitchen equipment, laboratory and medical equipment, and/or municipal steam systems. Existing appliances and equipment can be excluded.

AND/OR

OPTION 4: OUTDOOR WATER USE (1–2 POINTS)

PATH 1: NO IRRIGATION

Show that the landscape does not require a permanent irrigation system beyond a maximum **two-year establishment period**.

OR

PATH 2: EFFICIENT IRRIGATION

Reduce the project's theoretical irrigation requirement (TIR) by **at least 50%** from the calculated baseline. Points scale with the percent reduction (50%, 100%).

AND/OR

OPTION 5: OPTIMIZE PROCESS WATER USE (1–2 POINTS)

PATH 1: LIMIT COOLING TOWER CYCLES

Run a one-time water analysis to find the maximum safe number of cycles and limit cooling tower cycles to that level to earn 1 point. By treating the water for 25% more cycles or using 20% alternative water, project teams can instead earn 2 points.

OR

PATH 2: OPTIMIZE WATER USE FOR COOLING

This path applies only when the project's **ASHRAE 90.1 Appendix G** baseline system uses a cooling tower. For such projects, this path rewards using less water for cooling than that standard reference system would. To cut cooling water use, project teams can combine several strategies, such as maximizing the cycles of concentration, adding chemical treatment, installing drift eliminators, and using flow meters, water-level controls, and smart monitoring controls. For New Construction, a **25%** reduction earns 1 point and a 50% reduction earns 2.

OR

PATH 3: PROCESS WATER USE

Use alternative water to meet process water demand, with points scaling by the share of alternative water used (a minimum of 20%, or a minimum of 30%). Ensure that alternative water is of sufficient quality for its intended end use. Eligible process water uses must represent at least 10% of total building regulated water use and may not include water used for cooling.

AND/OR

OPTION 6: WATER REUSE (1–2 POINTS)

PATH 1: REUSE-READY SYSTEM

Install a water supply system that allows reclaimed or alternative water to reach one or more eligible end uses, providing space for treatment equipment as applicable.

OR

PATH 2: ALTERNATIVE WATER SOURCES

Implement water reuse strategies on-site and/or use reclaimed or alternative water supplied by municipalities.

OPERATIONAL CARBON PROJECTION AND DECARBONIZATION PLAN—PREREQUISITE

Complete all of the following.

- **Design Analysis** — analyze efficiency, peak load reduction, and decarbonization measures during the early design stages and account for the results in design decisions, using at least one of the following methodologies: simplified energy modeling, analysis from similar projects, or analysis from published data.
- **Site Energy Estimate** — estimate the amount of each type of energy the project will use annually in terms of site energy and submit the data to USGBC.
- **Review Carbon Projection** — using the submitted annual energy data, the project's current grid data, and the project's location, USGBC generates a business-as-usual (BAU) projection of the project's carbon emissions from energy use from the present through a **25-year** period (the projection assumes every grid decarbonizes by 95% over that period). Projects subject to a carbon-based building performance standard (BPS) must instead create an ordinance-specific BAU based on the electrical coefficients defined in the ordinance, with an overlay showing the applicable caps and, where applicable, the annual and cumulative fines or fees for exceeding the caps **over 25 years**. The building owner or owner's representative attests that they have reviewed the BAU carbon and fee projections.
- **Decarbonization Plan** — satisfy one of two paths. Under Path 1: Design for Electrification, earn 4 or 5 points in the Electrification credit. Under Path 2: Plan for Decarbonization, create a narrative of no more than two pages detailing how decarbonization could be achieved over a **25-year** period: the retrofits to be made with the approximate timeline and cost of each, the equipment or building materials to be discarded and the new equipment to be purchased, and the electrification-readiness strategies built into the initial design with a rough estimate of the avoided cost, disruption, and materials waste of each. Common readiness strategies include oversizing electrical panels and service, installing conduit for future loads, an enhanced envelope, or heating distribution that can accommodate the lower temperatures of future heat pumps. The owner or owner's representative attests that they have reviewed the plan.

MINIMUM ENERGY EFFICIENCY—PREREQUISITE

Comply with one version of ASHRAE 90.1 using any applicable compliance path in Section 4.2. Projects registering before January 1, 2028, may comply with either option; projects registering on or after January 1, 2028, must comply with Option 2.

OPTION 1: ASHRAE 90.1-2019

Comply with **ANSI/ASHRAE/IES Standard 90.1-2019** with Addendum cr, using any applicable compliance path in Section 4.2.

The team then picks one of three compliance paths, which are the prescriptive method, the Energy Cost Budget (ECB) method, and the Appendix G Performance Rating Method (PRM).

The prescriptive method requires meeting every requirement in each section of the standard on its own, covering the envelope, HVAC, service water heating, power, lighting, and equipment. This path does not allow trade-offs between systems.

The Energy Cost Budget (ECB) method, on the other hand, uses whole-building energy modeling. This path allows trade-offs between systems, since the model judges the building as a whole rather than system by system.

The Appendix G Performance Rating Method (PRM) also uses whole-building modeling, comparing the design against a fixed baseline building. It offers the most flexibility for trading off between systems, and teams pursuing the Enhanced Energy Efficiency credit must use it or the prescriptive method. For projects applying the Normative Appendix G Performance Rating Method, the future source energy metric may be used in place of cost: replace all references to cost with future source energy, using an electric site-to-source energy conversion factor of 2.0 based on future U.S. projections (a lower national average may be used for projects outside the U.S.), and replace the standard's Building Performance Factors (BPFs) with the BPFs derived for the future source energy metric.

OR

OPTION 2: ASHRAE 90.1-2022

Comply with **ANSI/ASHRAE/IES Standard 90.1-2022**. The team can use the same three compliance paths as Option 1, which are the prescriptive method, the Energy Cost Budget method, and the Appendix G Performance Rating Method.

FUNDAMENTAL COMMISSIONING—PREREQUISITE

Comply with the **ANSI/ASHRAE/IES Standard 90.1** commissioning requirements for building systems, controls, and the building envelope, with the additional provisions below. Use the 2019 version or later (errata included) for projects registered before January 1, 2028, or the 2022 version or later for projects registered on or after that date.

- All projects must provide fundamental commissioning.
- By the end of the design development phase, the owner designates a commissioning provider (CxP) with experience completing commissioning on at least two projects of equal or larger scope and complexity.
- In addition to the ASHRAE 90.1 requirements, the CxP develops the owner's project requirements (OPR) in predesign or as early as possible and reviews and updates it through design and construction; the OPR must include HVAC, service water heating, power, lighting, other equipment (including on-site renewable energy), and envelope.
- During design, the CxP reviews the basis of design (BOD) for compliance with the OPR and attends at least one meeting focused on mechanical, electrical, and plumbing and one focused on envelope (which may be separate or combined).
- During construction, the CxP reviews submittals and substitutions for design deviations that impact the OPR, attends milestone meetings at **50%** and **100%** completion, and performs a sample review of **at least 10%** of completed contractor documentation for quality assurance and quality control; for envelope, testing is included in the commissioning documents and the CxP witnesses a sample of tests (not required for Core and Shell projects).
- During occupancy and operations, the CxP develops an ongoing commissioning plan.

ENERGY METERING AND REPORTING—PREREQUISITE

Complete both of the following.

- Energy Monitoring and Recording — install or use existing devices to monitor and record energy use per **ANSI/ASHRAE/IES Standard 90.1** (the 2019 version or later for projects registered before January 1, 2028, or the 2022 version or later on or after that date), and install or use existing devices to monitor and record on-site renewable electricity generation, meeting the same monitoring and reporting requirements the standard sets for electrical end uses. Major renovations and buildings eligible for exceptions to ASHRAE 90.1-2019 must install devices capable of monitoring whole-building energy use for each energy source and building peak electricity demand at least monthly.
- Report Energy Data — commit to reporting to USGBC at least annually the monthly energy data for **12 consecutive months** of total consumption for each energy source, on-site renewable energy generation, and peak electrical demand. This commitment carries forward for **five years** or until the building changes ownership or lessee.

Core and Shell projects may except future tenant utility services and meters that will be installed in the tenant scope of work.

FUNDAMENTAL REFRIGERANT MANAGEMENT—PREREQUISITE

OPTION 1: NO REFRIGERANTS

- Do not use refrigerants in the project.

OR

OPTION 2: REFRIGERANTS

- Refrigerant inventory — inventory the refrigerant-containing equipment installed within the project scope and any existing equipment owned by the building owner, including the refrigerant type, GWP, the amount of refrigerant in each, and the total GWP of all refrigerants.
- Phase out high-GWP and HCFC refrigerants — do not use hydrochlorofluorocarbon refrigerants in new equipment, and evaluate available alternatives during design for any refrigerant with a GWP greater than 700.
- Leak check and repair — prior to substantial completion, check both new and existing refrigerant-containing equipment for leaks and repair all identified leaks. For systems with field-assembled joints, perform a leak check, vacuum check, and pressure check prior to charging with refrigerant.

ELECTRIFICATION—CREDIT (1–5 POINTS)

OPTION 1: NO ON-SITE COMBUSTION (5 POINTS)

- Design and operate the project from start-up with no on-site combustion except for emergency support systems.
- The combined weighted average equipment efficiency for space heating and service water heating (SWH) must be at least 1.8 COP.

- The following may be excluded from the COP determination: space-heating equipment in climate zones 0–2; supplemental heating equipment designed only for low-temperature operation; and SWH equipment in nonresidential spaces complying with the point-of-use water heater criteria in ASHRAE 90.1-2022 (W05) without exceptions.

OR

OPTION 2: NO ON-SITE COMBUSTION EXCEPT AT LOW TEMPERATURES (1–4 POINTS)

PATH 1: SPACE HEATING (2 POINTS)

- Design space heating to be capable of operating without on-site combustion except at low temperatures. Projects in climate zones 3 and above should achieve a weighted average space heating efficiency of at least 1.8 COP. Projects that install no space heating equipment and require none for occupant comfort earn this path automatically.

AND/OR

PATH 2: SERVICE WATER HEATING (1 POINT)

- Design SWH systems to be capable of operating without on-site combustion except at low temperatures.

AND/OR

PATH 3: COOKING AND OTHER PROCESS LOADS (1 POINT)

- Design cooking, laundry, process equipment, and on-site power generation (except emergency support systems) to be capable of operating without on-site combustion. Projects without these systems earn this point automatically; low-temperature process heating equipment may be excluded.

REDUCE PEAK THERMAL LOADS—CREDIT (1–5 POINTS)

Comply with any combination of Options 1–4. For all options, the building envelope must meet **ASHRAE 90.1** Section 5.5 (Prescriptive) or Section 5.6 (Trade-Off) per the version referenced in Minimum Energy Efficiency, and envelope efficiency may not be traded off against other building systems.

OPTION 1: INFILTRATION AND BALANCED VENTILATION (1 POINT)

- Balanced ventilation — design ventilation and exhaust airflows within 10% of each other, with a testing, adjusting, and balancing (TAB) report in the commissioning scope (does not apply to Core and Shell).
- Infiltration — demonstrate measured building-envelope air leakage at or below the LEED caps using an approved air leakage test (ASTM E779, ANSI/RESNET/ICC 380, ASTM E3158, ASTM E1827, or equivalent); buildings smaller than 25,000 square feet (2,322 square meters) must use a whole-building test. Residential projects instead compartmentalize and blower-door test dwelling units (ANSI/RESNET/ICC 380) to at most 1.5 times the caps per unit, with a compliant weighted average for the building.

AND/OR

OPTION 2: VENTILATION ENERGY RECOVERY (1 POINT)

- Each fan system supplying outdoor air must have an energy or heat recovery system with at least a 70% enthalpy recovery ratio or a 75% sensible heat recovery ratio, with provisions to bypass or control it during moderate outside-air conditions. Fan systems supplying less than 15% of the project's total outdoor air (in aggregate) may be excluded.

AND/OR

OPTION 3: THERMAL BRIDGING (1 POINT)

- Comply with the prescriptive thermal bridging requirements of **ASHRAE 90.1-2022** without applying the exceptions for projects in climate zones 0–3.

AND/OR

OPTION 4: PEAK THERMAL LOAD REDUCTIONS (1–3 POINTS)

- Include ventilation loads in the peak coincident loads, use measured envelope air leakage in the peak-load calculation, and demonstrate balanced ventilation per Option 1. Then pursue one of the following paths.

PATH 1: PEAK LOAD INTENSITY (1–3 POINTS)

- Limit the sum of peak heating and cooling load per unit of treated floor area (calculated with the WUFI Passive design tool or the Passive House Planning Package).

OR

PATH 2: ASHRAE 90.1 TRADE-OFF METHODS (1–3 POINTS)

- Pursue envelope and/or ventilation improvements: a 10% or 20% improvement in peak thermal loads from the envelope (ASHRAE 90.1-2022 Building Envelope Trade-Off, Normative Appendix C) earns 1 or 2 points, and/or a 10% improvement in peak coincident loads from ventilation (Mechanical System Performance Rating Method, Normative Appendix L) earns 1 point.

OR

PATH 3: ENERGY SIMULATION (1–3 POINTS)

- Demonstrate a performance index (per ASHRAE 90.1-2019 or later, Normative Appendix G, replacing all references to cost with the sum of building peak coincident heating and cooling loads) of 0.5, 0.4, or 0.3.

Residential: Option 1 earns 2 points; Options 2–4 follow the New Construction requirements.

ENHANCED ENERGY EFFICIENCY—CREDIT (1–10 POINTS)

OPTION 1: PRESCRIPTIVE PATH (1–10 POINTS)

Pursue Path 1 and/or Path 2.

PATH 1: REGULATED LOADS (1–7 POINTS)

Implement additional efficiency credits from **ASHRAE 90.1-2022**, **Section 11** (eligible measures: HVAC, service water heating, lighting, and building mass/night flush). Points scale with the number of Section 11 credits earned:

- Case 1 — **ASHRAE 90.1-2019** (only for projects registered before January 1, 2028): 1 point per 25 credits, up to 5 points at 125 credits.
- Case 2 — **ASHRAE 90.1-2022**: meeting the standard's minimum credits earns 4 points; each additional 25 credits earns 1 more, up to 7.

AND/OR

PATH 2: PLUG AND PROCESS LOADS (PPLs) (1–4 POINTS)

- Case 1 — Plug Load Management (1 point): Provide an app-based plug load dashboard to all regular occupants (tenants may opt out of displaying their loads) and, for building or tenant types with IT departments, control PCs, monitors, and displays off when not in use.

AND/OR

- Case 2 — Efficient Plug and Process Load Equipment (1–4 points): Install or reuse eligible equipment meeting the LEED criteria (ENERGY STAR plug and process loads, Class A people conveyance, efficient data-center electrical systems, refrigeration, etc.) for 90% of applicable equipment by quantity or rated load; points scale with the number of qualifying equipment categories or, for process-intensive buildings, with the compliant rated load.

OR

- Case 3 — Plug and Process Load Exceptional Calculation (1–4 points): Using the **ASHRAE 90.1 Section G2.5 Exceptional Calculation Method**, demonstrate a 10%, 20%, 30%, or 40% improvement in total plug, process, refrigeration, and conveyance loads.

OR

OPTION 2: ENERGY SIMULATION (1–10 POINTS)

- Demonstrate an improvement in future source energy per **ASHRAE 90.1 Normative Appendix G**, using the version applied in Minimum Energy Efficiency. Points scale with the percentage improvement in the performance index below its target — Path 1 (excluding on-site renewable contribution): 3% to 30%; Path 2 (including on-site renewable contribution): 10% to 100%.

RENEWABLE ENERGY—CREDIT (1–5 POINTS)

Supply or procure renewable energy from any combination of Tier 1, Tier 2, and/or Tier 3 sources; points documented for the tiers may be added together.

OPTION 1: RENEWABLE ENERGY SUPPLY OR PROCUREMENT (1–5 POINTS)

Points scale with the share of annual site energy from each tier.

- **Tier 1** — on-site renewable generation, campus generation, or a social-impact project (the renewable system is provided, installed, and commissioned at no cost to the entity, with ownership and power rights transferred to it). The eligible tier 1 resources are solar photovoltaics, solar thermal, wind, and recovered heat from municipal wastewater. Only the usable energy counts, which is the system output minus the transmission and conversion losses, such as the loss when converting DC electricity to AC.

- **Tier 2** — new off-site renewable electricity from generation assets contracted to be operational **within two years** of building occupancy, or contracted **no more than five years** after their commercial operation date.
- **Tier 3** — off-site renewable electricity or renewable fuels that are Green-e Energy certified or equivalent.
- Contracts must be **10 years** long (or prorated across 10 years for shorter contracts). All environmental attributes (EACs/RECs) must be retired on the project's behalf; EACs must be generated no earlier than **18 months** before the initial application; Tier 2 and Tier 3 assets must be in the same country or region as the project; and one-time or annual bulk purchases exceeding 100% of annual electricity use require **Green-e Energy certification** or equivalent.

AND/OR

OPTION 2: RENEWABLE ENERGY READINESS — CORE AND SHELL ONLY (1 POINT)

Design the project for renewable energy readiness and provide tenant guidelines showing how tenants can install renewable capacity. Document solar readiness (or equivalent on-site renewable readiness): a dedicated solar zone of at least 40% of gross roof area (optimally located and unshaded, no obstructions, with minimum dimensions), appropriate panel-mounting and roof-warranty provisions, reserved interconnection pathways for conduit and for inverters and metering, and an electrical service with at least a 200-amp busbar and reserved breaker space permanently marked 'For future solar electric.'

ENHANCED COMMISSIONING—CREDIT (1–4 POINTS)

OPTION 1: ENHANCED COMMISSIONING (1–3 POINTS)

The owner designates an independent commissioning provider (CxP) during **predesign or very early in the design phase**.

PATH 1: ENHANCED COMMISSIONING FOR MEP SYSTEMS (2 POINTS)

- Comply with **ANSI/ASHRAE/IES Standard 202-2024** (Commissioning Process) for mechanical, electrical, plumbing, control, data center, process, building-monitoring, and renewable energy systems. Additionally, during design the CxP attends **at least two coordination/design meetings**, and prior to or during occupancy reviews the training materials to confirm they meet the training plan and that training occurred.

AND/OR

PATH 2: ENHANCED COMMISSIONING FOR BUILDING ENCLOSURE (1 POINT)

- Comply with the tasks and deliverables of **ASTM E2947-21a** — building air leakage (ASTM E783, E779, E1186, or E3158), water penetration (ASTM E1105 or AAMA 501.2), and infrared imaging (ASTM C1153 or C1060) — and during occupancy confirm that the building enclosure commissioning (BECx) training occurred.

AND/OR

OPTION 2: MONITORING-BASED COMMISSIONING (MBCx) (1–2 POINTS)

PATH 1: BASIC SOFTWARE (1 POINT)

- Commit to **at least three years** of MBCx through a contract with an MBCx provider or qualified staff, commencing no later than building occupancy, with an MBCx plan covering roles and responsibilities, the software technology, and review and reporting criteria (staff training, prompt communication of major faults, at least quarterly anomaly summaries, at least annual reporting of trends, benchmarks, faults, and savings, and at least two system reviews).
- Provide a remotely accessible energy information system that presents all metered data from Energy Metering and Reporting, with annual energy-use-intensity benchmarking, interval comparisons, hourly electric load shapes, and submeter visualization, plus hourly monitoring of elevators, escalators, and moving walkways and of commercial kitchen and process equipment in spaces over 10 kW.

OR

PATH 2: ENHANCED SOFTWARE (2 POINTS)

- Comply with Path 1 and add fault detection and diagnostics (FDD) for projects whose installed cooling, heating, or refrigeration capacity exceeds 7,200 kBtu/hr (600 tons or 2,110 kW), covering at least 60% of capacity, with smart analytics, direct fault-to-trend linking, sorting and filtering, exportable reports, and at least three years of trend storage; enhance the energy information system with automated anomaly reporting, energy normalization, and greenhouse gas emissions reporting.

GRID INTERACTIVE—CREDIT (1–2 POINTS)

Evaluate grid-interactive measures against the current and forecasted grid context, location, building type, and ownership structure, and account for the results in decision-making. Provide interval recording meters and equipment capable of accepting an external signal. Pursue any combination of the options below.

OPTION 1: ENERGY STORAGE (1–2 POINTS)

- Provide on-site electric and/or thermal storage with automatic load-management controls that store energy during off-peak or low grid-carbon periods and use it during on-peak or high grid-carbon periods.

AND/OR

OPTION 2: DEMAND RESPONSE PROGRAM (1 POINT)

- Enroll in at least a **one-year** demand response (DR) contract with a qualified DR program provider, intending multiyear renewal. On-site electricity generation and fuel combustion cannot be used to meet the demand-side management criteria.

AND/OR

OPTION 3: AUTOMATED DEMAND-SIDE MANAGEMENT (1 POINT)

PATH 1: SYSTEM-LEVEL CONTROLS

- Provide automated demand response controls for **at least two** installed systems: HVAC (50% of rated capacity), lighting (50% of power), automatic receptacle controls, service water heating (90% of capacity), or electric vehicle supply equipment.

OR

PATH 2: BUILDING AUTOMATION SYSTEM

- Develop a plan to shed **at least 10%** of the project's peak electricity demand for **at least one hour**, addressing both winter and summer peaks under electrified-grid projections, and install a control system that automatically sheds demand in response to grid-strain or high-emissions triggers (a DR program signal, an API grid-emissions feed, a peak-demand tariff period, or a time-of-use price peak).

AND/OR

OPTION 4: POWER RESILIENCE (1 POINT)

- Identify critical equipment that requires continuous operation and design the project to island and operate independently from the grid, powering those critical loads with the project's on-site renewable and energy storage systems for **at least three days**.

ENHANCED REFRIGERANT MANAGEMENT—CREDIT (1–2 POINTS)

Pursue any combination of the options below. District-energy projects must comply at the district facility or follow additional guidance.

OPTION 1: NO REFRIGERANTS OR LOW GWP (1–2 POINTS)

PATH 1: NO REFRIGERANTS (1 POINT)

- Do not use refrigerant-containing equipment in the project. (Core and Shell: projects without future equipment necessary to meet the heating or cooling load are not eligible for this path.)

OR

PATH 2: LOW GWP REFRIGERANTS (1–2 POINTS)

- The project's total weighted average refrigerant GWP across all new refrigerant-containing equipment should not exceed **80%** of the weighted average of the GWP benchmarks for 1 point, or 50% for 2 points. The benchmarks are derived primarily from the EPA's 2023 AIM Act Technology Transitions Rule, in line with the EU F-Gas Regulations and the Kigali Amendment.

AND/OR

OPTION 2: LIMIT REFRIGERANT LEAKAGE (1 POINT)

- Design, construct, and operate the refrigerant-using equipment to minimize leakage: use self-contained equipment with no field-installed piping (for equipment over 700 GWP and for at least 80% of the total GWP of refrigerants used); specify automatic leak detection in fully enclosed spaces with a charge over 100 tCO_2e; use brazed or press-type fittings for any field-installed piping; and maintain a refrigerant management plan with a designated oversight party covering inventory updates, charge and leakage tracking, servicing by certified personnel, annual leak-detector calibration, periodic manual leak checks, defined repair timeframes, and doubled testing and repair frequency if annual recharge/leakage exceeds 1%.

AND/OR

OPTION 3: GREENCHILL CERTIFICATION FOR FOOD RETAILERS (1–2 POINTS)

- Available to projects where food retailing constitutes **more than 20%** of gross area. Achieve the **EPA GreenChill certification program** (or, internationally, the equivalent requirements): Silver earns 1 point; Gold or Platinum earns 2 points.

PLANNING FOR ZERO WASTE OPERATIONS—PREREQUISITE

Comply with both of the following.

- Storage and Collection of Recyclables — provide dedicated areas (which may be in separate locations) accessible to waste haulers, janitorial staff, and building occupants for the collection and storage of recyclable materials for the entire building. Recyclables must include organics/food waste, mixed paper, corrugated cardboard, glass, plastics, and metals; mixed recyclables are acceptable for paper, cardboard, glass, plastics, and metals where local conditions require, and space for organics/food waste is required even if service is not available at occupancy. Take appropriate measures for the safe collection, storage, and disposal of batteries, mercury-containing lamps, and electronic waste.
- Zero Waste Operations Planning — include design details, maintenance manuals, and/or other resources from the design and construction team that help building occupants and operators meet high-performance waste prevention and recycling goals once in operation.

Core and Shell projects communicate the building's waste infrastructure and service options in the tenant guidelines.

QUANTIFY AND ASSESS EMBODIED CARBON—PREREQUISITE

Comply with both of the following.

- Embodied Carbon: Quantify the embodied carbon impacts—global warming potential (GWP)—of the project's structure, enclosure, and hardscape materials. Include all ancillary structures within the LEED project boundary, such as parking structures or outbuildings. At a minimum, include the following material categories in the calculations: asphalt, concrete, masonry, structural steel, insulation, aluminum extrusions, structural wood and composites, cladding, and glass. Quantify cradle-to-gate (A1–A3) embodied carbon emissions for each material, calculated as the product's GWP per unit multiplied by the quantity used.
- High-Priority Embodied Carbon Sources: Identify the top three sources of embodied carbon on the project and describe how project-specific strategies were considered to reduce the impacts of these hot spots.

Core and Shell projects communicate the embodied carbon measurements and material suppliers in the tenant guidelines.

BUILDING AND MATERIALS REUSE—CREDIT (1–3 POINTS)

OPTION 1: BUILDING REUSE (1–3 POINTS)

Maintain the existing building structure — including floor decking, roof decking, and enclosure — and calculate reuse by project area (portions deemed structurally unsound or hazardous are excluded). The reused area includes both the elements kept in place and any salvaged structural elements procured from off-site and incorporated into the building. For projects with additions, the calculation is based on the existing floor area only—the new wing does not dilute the math. For New Construction, reusing 20%, 35%, or 50% of the existing structure and enclosure earns 1–3 points.

AND/OR

OPTION 2: MATERIALS REUSE (1–2 POINTS)

Survey and identify opportunities to reuse materials in place or acquire them from off-site salvage sources or reuse markets and incorporate them into the new design; for projects with deconstruction or demolition in scope, conduct a salvage assessment first (materials sent off-site for reuse instead contribute to Construction and Demolition Waste Diversion). Earn 2 points by reusing **at least 30%** of one targeted material type, **at least 15%** of two targeted material types, **at least 15%** of four other material types, or an equivalent weighted average. Targeted materials (valued higher) include carpeting, ceilings, furniture, and interior walls; other materials include dimensional lumber, doors, casework, floor coverings, lighting and plumbing fixtures, mechanical equipment, and door hardware.

REDUCE EMBODIED CARBON—CREDIT (1–6 POINTS)

Quantify the reduction of embodied carbon of the major structure, enclosure, and hardscape materials, including all ancillary structures within the LEED project boundary. Baseline and final results may use as-designed or as-constructed quantities provided quantities did not change **more than 10%** from design through construction.

OPTION 1: WHOLE-BUILDING LIFE-CYCLE ASSESSMENT (1–6 POINTS)

- Conduct a cradle-to-grave (modules A–C, excluding operating energy and operating water-related energy) whole-building life-cycle assessment of the structure, enclosure, and hardscape, comparing results to a baseline developed for the project and reporting the required impact categories: GWP, stratospheric ozone depletion, acidification, eutrophication, tropospheric ozone formation, and depletion of nonrenewable energy resources. Points scale with the GWP reduction against the baseline — from meeting the baseline or industry average up through reductions of 10%, 20%, 30%, and 40% or more.

AND/OR

OPTION 2: ENVIRONMENTAL PRODUCT DECLARATION (EPD) ANALYSIS (1–3 POINTS)

PATH 1: PROJECT-AVERAGE APPROACH (1–3 POINTS)

- Reduce the project's whole-project weighted-average embodied carbon relative to industry averages (U.S. EPA, the most recent Carbon Leadership Forum Material Baselines report, or similar) using product-specific or facility-specific Type III EPDs; points scale with the reduction achieved, from meeting the baseline up through 20% and 40%+ reductions.

OR

PATH 2: MATERIALS-TYPE APPROACH (1–2 POINTS)

- Demonstrate via product-specific Type III EPDs that targeted structural, enclosure, and hardscape material types have lower embodied carbon than industry benchmarks; points scale with the number of qualifying material categories — three or five categories for New Construction.

AND/OR

OPTION 3: TRACK CARBON EMISSIONS FROM CONSTRUCTION ACTIVITIES (1–2 POINTS)

- Track all fuel and utility usage (LCA module A5) for contractor jobsite operations (1 point), or for both contractor and subcontractor jobsite operations (2 points).

LOW-EMITTING MATERIALS—CREDIT (1–2 POINTS)

Specify and install permanently installed products — paints, coatings, adhesives, sealants, flooring, walls, ceilings, insulation, furniture, and/or composite wood — that meet the low-emitting criteria. Points are awarded by product category.

- Path 1 (1 point) — more than 90% of products (by volume, cost, or surface area) in each of three categories: paints and coatings, flooring, and ceilings.
- Path 2 (2 points) — achieve Path 1 plus more than 80% of each of any two additional categories: adhesives and sealants, walls, insulation, or composite wood.
- Path 3 (2 points) — achieve Path 1 plus more than 80% of the furniture category.

Compliance is demonstrated through the VOC emissions criteria (a qualifying third-party certification or independent laboratory report to the CDPH Standard Method v1.2-2017, or an inherently non-emitting, salvaged, or reused product), the furniture emissions criteria (ANSI/BIFMA M7.1 and e3), and the formaldehyde emissions criteria (CARB/EPA-recognized ULEF or no-added-formaldehyde certification). Fire suppression, HVAC, plumbing, electrical, conveying and communications equipment, poured concrete, structural framing, SIPs, and water-resistive barriers are not applicable.

For Core and Shell, products to be specified and installed by the tenants may be excluded from the calculations.

BUILDING PRODUCT SELECTION AND PROCUREMENT—CREDIT (1–5 POINTS)

Select nonstructural interior and enclosure building products that demonstrate achievement in one or more of five criteria areas — climate health, human health, ecosystem health, social health and equity, and circular economy — shown through eligible manufacturer product documentation (third-party certifications, ecolabels, declarations, and standards). These five criteria areas are designed to align with the **Mindful Materials Common Materials Framework** and the **American Institute of Architects (AIA) Materials Pledge**. Products achieving two or more criteria areas are multi-attribute.

- Each product earns an achievement level per criteria area: Level 1 (a first step toward sustainability) earns a 1× multiplier, Level 2 (a marketplace leadership position) a 2× multiplier, and Level 3 (elite, at the forefront of sustainability) a 3× multiplier. Scores are added across criteria areas up to a maximum multi-attribute score (MAS) of 5 per product, and each product's value (cost, area, volume, or unit) is multiplied by its MAS.
- Eligible product categories are paints and coatings, adhesives and sealants, flooring, walls, ceilings, insulation, furniture, composite wood, and plumbing fixtures. A product category whose multi-attribute adjusted value exceeds 100% of the category's total value earns 1 point (up to 5 points total).

CONSTRUCTION AND DEMOLITION WASTE DIVERSION—CREDIT (1–2 POINTS)

Comply with both of the following.

- Construction and Demolition Materials Management Plan — develop and implement a construction and demolition (C&D) materials management plan.
- Diversion — follow the plan and provide a final waste management report detailing all waste generated, with disposal and diversion rates (calculations by weight or volume, consistent throughout). Points scale with the diversion rate: divert at least 50% of the total C&D material (with at least 10% of diverted materials salvaged or source-separated and sent to single-material recyclers), or at least 75% (with at least 25% salvaged or source-separated).
- Source-separated materials count as 100% diverted, and salvaged materials sent off-site for reuse count at 200%. Mixed C&D materials sent to a processing facility take the facility average recycling rate (a maximum of 35% if not third-party verified). Materials sent to alternative daily cover or incineration/energy recovery count as waste (0% diverted). Exclude hazardous waste, excavated soil and land-clearing debris, and on-site reuse.

Core and Shell projects include the approved construction and demolition waste management plan in the tenant guidelines.

CONSTRUCTION MANAGEMENT—PREREQUISITE

Develop and implement construction management practices for the construction and preoccupancy phases that address all of the following.

- **No smoking** — prohibit smoking during construction except in designated smoking areas located **at least 25 feet (7.5 meters)** from the building, with signage prohibiting smoking during construction.
- **Extreme heat protection** — implement measures that protect construction workers from extreme heat.
- **HVAC protection** — keep contaminants out of the HVAC system; avoid running permanently installed equipment where possible, but if it must be used, maintain proper filtration; replace all air filtration media after construction and before occupancy; and confirm that testing and balancing is completed with the new filtration.
- **Source control** — keep sources of contaminants out of the building and plan to eliminate any that are introduced; store absorptive materials (carpets, acoustic ceiling panels, fabric wall coverings, insulation, upholstery, and furnishings) in a designated area protected from moisture damage.
- **Pathway interruption** — prevent circulation of contaminated air during activities such as cutting concrete or wood, sanding drywall, or installing VOC-emitting materials; isolate work areas with sealed doorways and windows or temporary barriers, use walk-off mats at entryways, and use dust guards and collectors on saws and other tools.
- **Housekeeping** — maintain a clean jobsite using vacuum cleaners with high-efficiency particulate filters and sweeping compounds or wetting agents for dust control.
- **Scheduling** — sequence construction activities to reduce air quality problems, and for renovation projects, coordinate activities to minimize or eliminate disruption of operations in occupied areas.

FUNDAMENTAL AIR QUALITY—PREREQUISITE

Comply with all of the following.

- Investigate Regional and Local Air Quality — investigate outdoor air quality in accordance with **ASHRAE Standard 62.1-2022**, Sections 4.1–4.3.
- Ventilation and Filtration Design — meet **ASHRAE 62.1-2022, Sections 5 and 6,** using the ventilation rate procedure, the IAQ procedure, the natural ventilation procedure, or a combination. Where the ventilation rate or IAQ procedure is used, each central HVAC system supplying outdoor or recirculated air to regularly occupied spaces must provide filtration of at least MERV 13 (ASHRAE 52.2-2017), an equivalent ePM1 50% media class (ISO 16890-2016), or tested in-room air cleaning systems; and provide outdoor airflow measurement devices for systems with intake flow greater than 1,000 cfm (472 L/s). Healthcare spaces meet ASHRAE 170-2021, Sections 6–10.
- Entryway System Design — install permanent entryway systems to capture dirt and particulates at primary exterior entrances (no length requirement). Although LEED does not enforce a minimum length requirement for entryway systems, LEED advises project teams to use at least **10 feet** (**3 meters**) in the primary direction of travel.

Residential dwelling units additionally require: a dwelling-unit mechanical ventilation system per **ASHRAE 62.2-2022** supplying outdoor air; local exhaust in each kitchen and bathroom exhausted outdoors; no unvented combustion appliances (ovens and ranges excepted); a hard-wired carbon monoxide monitor with battery backup on each floor; sealed glass enclosures and backdraft testing for indoor fireplaces and woodstoves; and closed-combustion or power-vented combustion space- and water-heating equipment.

NO SMOKING OR VEHICLE IDLING—PREREQUISITE

Comply with both of the following, and communicate the policy to occupants with provisions for enforcement or prohibitive signage.

- Prohibit smoking — prohibit smoking inside the building (with limited exceptions) and outside except in designated smoking areas located **at least 25 feet (7.5 meters)**, or the maximum extent allowable by local codes, from all entries, outdoor air intakes, and operable windows. School projects prohibit all smoking on-site.
- Prohibit vehicle idling — prohibit vehicle idling on-site.

Residential projects meet these requirements for all areas except dwelling units and private balconies. If a residential project allows smoking inside dwelling units, then any unit where smoking is permitted must meet compartmentalization requirements by performing a blower door test in accordance with **ANSI/RESNET/ICC 380** (or equivalent).

ENHANCED AIR QUALITY—CREDIT (1 POINT)

Design the building to exceed **ASHRAE 62.1-2022, Section 6**. If the ventilation rate procedure was used for Fundamental Air Quality, use Option 1 or Option 2; if the IAQ procedure was used, use Option 2. Both options use the measures described in **ASHRAE Guideline 42-2023**, Enhanced Indoor Air Quality in Commercial and Institutional Buildings, the design guide for enhanced indoor air quality.

OPTION 1: INCREASED VENTILATION (1 POINT)

- Increase breathing-zone outdoor air ventilation rates **at least 15%** above the minimum rates determined in Fundamental Air Quality (30% for exemplary performance), providing the increased rates to **95%** of all regularly occupied spaces.

OR

OPTION 2: ENHANCED INDOOR AIR QUALITY DESIGN (1 POINT)

- In addition to the design compounds and limits in ASHRAE 62.1-2022, Tables 6-5 and 6-6, design and verify enhanced IAQ meeting the LEED's lower design limits: PM2.5 at 10 $\mu g/m^3$, formaldehyde at 20 $\mu g/m^3$, and ozone at 10 ppb.

OCCUPANT EXPERIENCE—CREDIT (1–7 POINTS)

OPTION 1: BIOPHILIC ENVIRONMENT (1–4 POINTS)

PATH 1: INTEGRATED BIOPHILIC DESIGN (1 POINT)

- Integrate biophilic design demonstrating the five principles adapted from Kellert and Calabrese: repeated and sustained engagement with nature; a focus on human adaptations to the natural world; emotional attachment to the building and its location; positive people–nature interactions; and mutually reinforcing, integrated architectural solutions.

AND/OR

PATH 2: QUALITY VIEWS (2–3 POINTS)

- Provide a view to the outdoor natural or urban environment for **75%** or 90% of all regularly occupied floor area, through glass with visible light transmittance above 40%, with the view including nature, urban landmarks, art, or objects **at least 25 feet (7.5 meters)** away, and with occupants having direct access within three times the head height of the glazing (interior atria views may meet up to 30%).

AND/OR

OPTION 2: ADAPTABLE ENVIRONMENT (1 POINT)

- Give occupants choice and flexibility to adapt the space, with variability in the thermal, sound, and lighting environments and the ability to move between sensory zones, including at least one accessible quiet space, plus at least one additional strategy — prospect-and-refuge seating with alternate perimeter paths, varied furniture and seating, height variety or height-adjustable desks, or accessible flexible outdoor/transitional space within **200 feet** (**60 meters**) of a building entrance.

AND/OR

OPTION 3: THERMAL ENVIRONMENT (1 POINT)

- Design indoor occupied spaces to meet **ASHRAE Standard 55-2023** (with errata) and explain how the design addresses thermal conditions that adjust with the changing seasons, overcooling during nontemperate seasons, and support for occupants performing tasks with varying levels of movement.

AND/OR

OPTION 4: SOUND ENVIRONMENT (1–2 POINTS)

PATH 1: MAPPING ACOUSTIC EXPECTATIONS FOR INDOOR AND OUTDOOR SPACES (1 POINT)

- Early in design, map the acoustic expectations (noise exposure, acoustic comfort, acoustic privacy, communication, and soundscape) for each primary indoor and outdoor space and define acoustic criteria and design strategies to meet them.

OR

PATH 2: ACOUSTIC CRITERIA FOR INDOOR AND OUTDOOR SPACES (2 POINTS)

- Demonstrate through calculations, modeling, and/or measurements that the mapping exercise informed design solutions meeting the acoustic criteria for **at least 75%** of occupied spaces and all classrooms and core learning spaces.

AND/OR

OPTION 5: LIGHTING ENVIRONMENT (1–6 POINTS)

PATH 1: SOLAR GLARE (1 POINT)

- Provide manual or automatic (with manual override) glare-control devices in all regularly occupied spaces that receive direct or reflected sun (spaces designed for direct sunlight may be excluded).

AND/OR

PATH 2: QUALITY ELECTRIC LIGHTING (1 POINT)

- Control luminaire glare (calculated luminance below 6,000 cd/sq. m. between 45 and 90 degrees from nadir, or a unified glare rating of 19 or lower) and provide a color rendering index of at least 90 or the equivalent IES TM-30-20 color rendering.

AND/OR

PATH 3: PROXIMITY TO WINDOWS FOR DAYLIGHT ACCESS (1 POINT)

- Locate at least 30% of the regularly occupied area within a 20-foot (6-meter) horizontal distance of envelope glazing with visible light transmittance above 40%.

AND/OR

PATH 4: DAYLIGHT SIMULATION (1–4 POINTS)

- Through an IES LM-83-23 daylight simulation, achieve an average spatial daylight autonomy (sDA300/50%) of at least 40%, 55%, 65%, or 75%, addressing glare in spaces where annual sunlight exposure exceeds 20%.

Core and Shell (1–7 points): the credit offers Option 1: Biophilic Environment (1–4 points, with Integrated Biophilic Design, Quality Views, and an added Outdoor Connections path for 1 point) and Option 2: Lighting Environment (1–6 points, with the same solar glare, quality electric lighting, proximity to windows, and daylight simulation paths).

ACCESSIBILITY AND INCLUSION—CREDIT (1 POINT)

Support access for people with physical disabilities by meeting all locally applicable accessibility codes identified in the Human Impact Assessment prerequisite; where no code is in place, provide ramps on accessible routes and regularly used exterior entrances, a minimum 32-inch (0.86-meter) clear width at doors meant for human passage, and wheelchair-accessible front-approach reception, security, and service counters. Then include at least 10 of the following strategies (those most relevant to the project) that go beyond the local code.

- **Physical diversity** — wave-to-open or hand/foot-press door operators at regularly used entrances, meeting spaces accommodating mobility devices for **at least 10%** of occupants, accessible fitness equipment with accessible routes, and an alternate accessible route wherever an inaccessible route such as stairs is provided.

- **Safety and aging** — nonslip flooring, secured area rugs with transition strips, visual indication or railing at full-height glazing, audible and visual emergency alerts, closed stair risers, visual contrast between walls, floors, doors, and casework, and visual, tactile, or photoluminescent warnings at floor-level changes.
- **Social health** — lactation room pods, at least one accessible all-gender single-use restroom or one multiuse all-gender restroom per floor, an adult changing station in a designated accessible or family restroom, signage in all languages spoken by **more than 5%** of the local population, and support for neurodivergent users through integrated biophilic design.
- **Navigation** — wayfinding signage clearly indicating exits, entrances, and major functions; nontext diagrams and symbols; braille, visual and auditory cues, or continuous linear indicators on paths of travel; patterns and color blocking to identify key spaces; and haptic/tactile wayfinding maps.

RESILIENT SPACES—CREDIT (1–2 POINTS)

OPTION 1: MANAGEMENT MODE FOR EPISODIC OUTDOOR AMBIENT CONDITIONS — NEW CONSTRUCTION ONLY (1 POINT)

- Design systems with the capability to operate an episodic outdoor-event management mode as described in ASHRAE Guideline 44 (addressing conditions such as wildland fire smoke), include the mode in the design and commissioning documents, and verify proper implementation during commissioning.

AND/OR

OPTION 2: MANAGEMENT MODE FOR RESPIRATORY DISEASES — NEW CONSTRUCTION ONLY (1 POINT)

- Design occupied spaces with the capability to operate an infection-risk management mode providing the minimum equivalent clean airflow rates of ASHRAE 241-2023, Section 5.1, include the mode in the design and commissioning documents, and verify proper implementation during commissioning.

AND/OR

OPTION 3: DESIGN FOR OCCUPANT THERMAL SAFETY DURING POWER OUTAGES (1–2 POINTS)

PATH 1: CONSIDER EXTREME HEAT (1 POINT)

- Demonstrate through thermal modeling that the building passively maintains thermally habitable conditions during a two-day power outage in peak summertime conditions of a TMY, with designated thermal safety zones.

AND/OR

PATH 2: CONSIDER EXTREME COLD (1 POINT)

- Demonstrate through thermal modeling or Passive House certification that the building passively maintains thermally habitable conditions during a two-day power outage in peak wintertime conditions, with designated thermal safety zones.

AND/OR

OPTION 4: OPERABLE WINDOWS (1–2 POINTS)

- Design **50%** (1 point) or **75%** (2 points) of the regularly occupied spaces to have operable windows capable of providing access to outdoor air during heat waves or localized power outages, meeting the opening size and location requirements of ASHRAE 62.1-2022, Section 6.4.

AIR QUALITY TESTING AND MONITORING—CREDIT (1–2 POINTS)

New Construction only. Pursue Option 1 and/or Option 2.

OPTION 1: PREOCCUPANCY AIR TESTING (1–2 POINTS)

- After construction ends and before occupancy (within 14 days of occupancy for retail projects), under ventilation conditions typical for occupancy, conduct baseline IAQ testing in representative locations, with the number of measurements scaled to the total occupied floor area.

PATH 1: PARTICULATE MATTER AND INORGANIC GASES (1 POINT)

- Over a four-hour period, test that carbon monoxide (peak concentration), ozone, PM2.5, and PM10 (average concentration) do not exceed the LEED concentration limits, using an allowed test method.

AND/OR

PATH 2: VOLATILE ORGANIC COMPOUNDS (1 POINT)

- Perform a TVOC screening test; if TVOC exceeds 500 $\mu g/m^3$, investigate individual VOCs against health-based limits, correct any issues, and retest; and demonstrate that individual VOCs (formaldehyde, benzene, toluene, and others) do not exceed the LEED limits, using ISO/IEC 17025-accredited laboratories.

AND/OR

OPTION 2: CONTINUOUS INDOOR AIR MONITORING (1 Point)

- Provide building-grade or better indoor air monitors, located **3–6 feet** (1–2 meters) above the floor, for carbon dioxide, PM2.5, total volatile organic compounds, temperature, and relative humidity.

PROJECT PRIORITIES—CREDIT (1–9 POINTS)

Achieve any combination of the following.

- **Regional Priority** — achieve a regional priority credit from USGBC's Project Priority Library, identified by USGBC as having additional regional importance for the project's region.
- **Project-Type Credits** — achieve a project-type credit from USGBC's Project Priority Library, identified as addressing unique needs for the given adaptation or building application.
- **Exemplary Performance** — achieve an exemplary performance credit from USGBC's Project Priority Library, identified as going above and beyond an existing LEED v5 prerequisite or credit in the priority areas of scale, decarbonization, resilience, health, equity, and/or ecosystems.
- **Pilot Credits** — achieve a pilot credit from USGBC's Project Priority Library.
- **Innovative Strategies** — achieve significant, measurable environmental performance using a strategy not addressed in the LEED v5 green building rating system, identifying the intent of the proposed innovation strategy, the proposed requirements for compliance, the proposed submittals to demonstrate compliance, and the design approach or strategies used to meet the requirements.

LEED AP—CREDIT (1 POINT)

At least one principal participant of the project team must be a LEED AP with a specialty appropriate for the project.

SUMMARY OF ASHRAE STANDARDS

Name of the Standard / Program	Keywords	Related Prerequisites / Credits
ASHRAE 52.2-2017	Filtration requirements (minimum efficiency reporting value, MERV)	Prerequisite—Fundamental Air Quality
ASHRAE 55-2023	Thermal comfort requirements	Credit—Occupant Experience
ASHRAE Standard 62.1-2022	Ventilation and outdoor air quality requirements	Prerequisite—Fundamental Air Quality, Credit—Enhanced Air Quality, Credit—Reduce Peak Thermal Loads, Credit—Resilient Spaces
ASHRAE 62.2-2022	Residential ventilation and exhaust requirements	Prerequisite—Fundamental Air Quality
ANSI/ASHRAE/IES Standard 90.1 (2019 and 2022)	Energy efficiency, commissioning, energy metering, thermal bridging, peak load reduction, energy simulation, and energy performance improvement	Prerequisite—Minimum Energy Efficiency, Prerequisite—Fundamental Commissioning, Prerequisite—Energy Metering and Reporting, Credit—Reduce Peak Thermal Loads, Credit—Enhanced Energy Efficiency
ASHRAE 90.4-2022	Data center energy and electrical system efficiency	Credit—Enhanced Energy Efficiency
ASHRAE Standard 202-2024	Commissioning process	Credit—Enhanced Commissioning
ASHRAE 241-2023	Air cleaning and infection risk management	Prerequisite—Fundamental Air Quality, Credit—Resilient Spaces
ASHRAE Guideline 42-2023	Enhanced indoor air quality measures	Credit—Enhanced Air Quality
ASHRAE Guideline 44	Protecting occupants from smoke during wildfire events	Credit—Resilient Spaces

IMPORTANT STANDARDS AND PROGRAMS

Name of the Standard / Program	Keywords	Related Prerequisites / Credits
ANSI Consensus National Standard Guide 2.0 for Design and Construction of Sustainable Buildings and Communities	Consensus methodology for running an integrative, cross-disciplinary design process and discovery/charrette workflow	Credit—Integrative Design Process
IPCC Sixth Assessment Report and Shared Socioeconomic Pathway (SSP) emissions scenarios	Climate science source; defines the SSP emissions scenarios (SSP1-1.9 to SSP5-8.5) used to project future hazards and design service life	Prerequisite—Climate Resilience Assessment, Prerequisite—Carbon Assessment
FEMA National Risk Index	Maps natural-hazard exposure/risk by location to identify project vulnerabilities	Prerequisite—Climate Resilience Assessment, Prerequisite—Human Impact Assessment
NOAA Climate Explorer	Historical data and county-level climate projections for hazard assessment	Prerequisite—Climate Resilience Assessment
ASCE/SEI 7 (Minimum Design Loads, incl. ASCE Hazard Tool and Windspeed tool)	Defines design wind/seismic loads, hurricane-prone regions, and Risk Categories; used for resilient site/structure design	Prerequisite—Climate Resilience Assessment, Credit—Enhanced Resilient Site Design
U.S. Department of Agriculture (USDA) Natural Resources Conservation Service	Defines prime farmland, unique farmland, and farmland of statewide importance	Credit—Sensitive Land Protection
U.S. Endangered Species Act / U.S. Fish and Wildlife Service	Defines endangered and threatened species and their critical habitat	Credit—Sensitive Land Protection
NatureServe	Assigns conservation status ranks for species and ecosystems	Credit—Sensitive Land Protection

Name of the Standard / Program	Keywords	Related Prerequisites / Credits
IUCN Red List of Threatened Species	Identifies endangered and threatened species for international projects	Credit—Sensitive Land Protection
Institute of Transportation Engineers (ITE) Parking Generation Manual	Provides base parking capacity ratios	Credit—Transportation Demand Management
National Electrical Code (NFPA 70)	Electrical installation requirements for electric vehicle supply equipment (EVSE)	Credit—Electric Vehicles
U.S. EPA Construction General Permit (CGP)	Sets the erosion and sedimentation control (ESC) plan requirements for construction activity	Prerequisite—Minimized Site Disturbance
Sustainable SITES Initiative (SITES v2)	Sustainable landscape and site-design credits for soils, vegetation, resilience, and wildfire risk reduction	Credit—Minimized Site Disturbance, Credit—Biodiverse Habitat, Credit—Accessible Outdoor Space, Credit—Enhanced Resilient Site Design
American Bird Conservancy Threat Factor Database	Defines bird-friendly (bird-collision-deterrent) glazing and material requirements	Credit—Biodiverse Habitat
ASCE 24 (Flood Resistant Design) and FEMA 543	Flood-resistant design of critical utilities, structures, and materials	Credit—Enhanced Resilient Site Design
FORTIFIED Commercial (High Wind and Hail)	Hail- and wind-resilient design requirements	Credit—Enhanced Resilient Site Design
U.S. National Tsunami Hazard Mitigation Program / TsunamiReady Guidelines	Tsunami-resilient site planning, signage, and evacuation	Credit—Enhanced Resilient Site Design
National Wildfire Coordinating Group (NWCG) Standards for Mitigation / Safer from the Start	Wildfire protection and defensible-space (fuel-reduction) measures	Credit—Enhanced Resilient Site Design
American Forests (Tree Equity Score)	Measures tree-canopy distribution and equity	Credit—Heat Island Reduction

Name of the Standard / Program	Keywords	Related Prerequisites / Credits
Illuminating Engineering Society (IES) BUG rating method / International Dark-Sky Association	Backlight-uplight-glare luminaire classification for light-pollution control	Credit—Light Pollution Reduction
WaterSense (U.S. EPA)	Identifies high-performance, water-efficient fixtures and fittings; theoretical irrigation requirement (TIR) methodology	Prerequisite—Minimum Water Efficiency, Credit—Enhanced Water Efficiency
eGRID (U.S. EPA Emissions & Generation Resource Integrated Database)	Subregional grid emissions factors used for the business-as-usual (BAU) carbon projection	Prerequisite—Operational Carbon Projection and Decarbonization Plan
ENERGY STAR	Certification for EV supply equipment, water-using appliances, and efficient plug and process equipment	Credit—Electric Vehicles, Prerequisite—Minimum Water Efficiency, Credit—Enhanced Water Efficiency, Credit—Enhanced Energy Efficiency
ISO 25745	Energy-performance classification for elevators, escalators, and moving walkways	Credit—Enhanced Energy Efficiency
ANSI/RESNET/ICC 380 and ASTM E779 / E3158 / E1827	Building-envelope air-leakage (blower-door) testing methods	Credit—Reduce Peak Thermal Loads
Passive House Institute (PHIUS) — WUFI Passive / Passive House Planning Package	Peak heating and cooling load modeling and passive thermal survivability	Credit—Reduce Peak Thermal Loads, Credit—Resilient Spaces
Green-e Energy	Certification for renewable electricity and energy attribute certificates (EACs/RECs)	Credit—Renewable Energy
U.S. EPA GreenChill	Refrigerant-management certification program for food retailers	Credit—Enhanced Refrigerant Management
Environmental Product Declaration (EPD) — Type III (ISO 14025 / ISO 21930 / EN 15804)	Third-party product life-cycle (global warming potential) disclosure	Prerequisite—Quantify and Assess Embodied Carbon, Credit—Reduce Embodied Carbon, Credit—Building Product Selection and Procurement

Name of the Standard / Program	Keywords	Related Prerequisites / Credits
Carbon Leadership Forum (CLF) Material Baselines	Industry-average embodied-carbon benchmarks	Prerequisite—Quantify and Assess Embodied Carbon, Credit—Reduce Embodied Carbon
Declare label and equivalent multi-attribute ecolabels	Product transparency, optimization, and circular-economy disclosures	Credit—Building Product Selection and Procurement
ISO 16890-2016	Particulate air-filter classification (ePM1) equivalent to MERV filtration	Prerequisite—Fundamental Air Quality
California Department of Public Health (CDPH) Standard Method v1.2-2017	VOC emissions testing criteria for materials and indoor air	Credit—Low-Emitting Materials, Credit—Air Quality Testing and Monitoring
ANSI/BIFMA (M7.1 and e3)	Furniture emissions evaluation	Credit—Low-Emitting Materials
CARB ATCM / U.S. EPA TSCA Title VI	Composite-wood formaldehyde emission limits (ULEF / no-added-formaldehyde)	Credit—Low-Emitting Materials
SMACNA IAQ Guidelines for Occupied Buildings Under Construction	Construction-phase indoor-air-quality control measures	Prerequisite—Construction Management
IES TM-30-20 and LM-83-23	Color rendering (TM-30) and daylight simulation — spatial daylight autonomy / annual sunlight exposure (LM-83)	Credit—Occupant Experience
NEMA Unified Glare Rating (UGR) white paper	Luminaire glare evaluation	Credit—Occupant Experience
ISO/IEC 17025	Testing-laboratory accreditation for indoor-air-quality testing	Credit—Air Quality Testing and Monitoring

CONCLUSION

First of all, I would like to congratulate you on finishing this book and personally thank you for choosing it for your exam preparation. I truly hope it has exceeded your expectations.

To truly evaluate and sustain our performance, we all need feedback. With that in mind, please share any comments about this book, your exam score, and/or any other opinions you would like to share with me. If you found this book helpful, spreading the word by leaving a review on your favorite book review site, such as Amazon, Goodreads, or others, would mean the world to me. Your review supports my work and helps guide other readers.

If you'd like to receive my latest articles, books, and courses, be sure to join my mailing list.

Thank you for being part of this journey. I wish you all the best in your exam and your career!

Scan the following QR codes to:	
Send your feedback to the author and/or subscribe to the author's mailing list	**Follow the author on LinkedIn**

GLOSSARY

40/60 Rule: A method used to choose the appropriate rating system for the project if the project seems to fit under multiple rating systems.

Active Daylighting: A system that tracks and collects sunlight using mechanical devices.

Active Transportation: Human-powered mobility such as biking, walking, or wheeling.

Adapted Plants: Types of plants that do not occur naturally in a specific location; however, they can nonetheless adapt easily to the climate of the region.

Adaptive Reuse (**Designing for Flexibility**): A design strategy that allows buildings to accommodate future changes in layouts or floor plans without major reconstruction, particularly useful in buildings that may need frequent modifications.

Addenda: The overarching term that encompasses all changes, improvements, and modifications made to a LEED rating system. Addenda includes four types of changes: corrections, clarifications, LEED interpretations, and alternative compliance paths (ACPs).

Advisory Task Forces: Short-duration groups designated by USGBC staff to informally advise on issues limited in scope during the LEED development process.

Albedo: A measure of surface reflectivity on a scale from 0 to 1, where 0 represents a black surface that absorbs all solar radiation and 1 represents a white surface that reflects all solar radiation.

Alternative Compliance Paths (ACPs): Provide additional options for meeting credit requirements for international projects that need to reference local or regional standards instead of United States–based standards. There are two types of ACPs: Regional ACPs, which are credit requirements that may optionally be applied in specific regions of the world, and pilot ACPs, which are paths USGBC is considering for inclusion in future versions of LEED and can optionally be used in place of existing LEED credits.

Alternative Fuel Vehicles: Vehicles that consume nongasoline, low-polluting fuels like hydrogen, electricity, propane, compressed natural gas, liquid natural gas, methanol, or ethanol.

Alternative Fuel: Low-polluting fuels like hydrogen, electricity, propane, compressed natural gas, liquid natural gas, methanol, or ethanol.

Alternative Water Source: Nonpotable water from on-site surfaces, or freshwater sources, such as graywater, on-site reclaimed water, collected rainwater, captured condensate, and rejected water from reverse osmosis systems. Water from public utilities is excluded.

Arc: The online platform where project teams register their projects for LEED certification, complete prerequisite and credit documentation, and submit applications to GBCI for certification review. It serves as the primary tool for managing the entire LEED certification process from start to finish.

Basis of Design (BOD): Describes the information necessary to accomplish the owner's project requirements, including system requirements, design criteria, standards, guidelines, etc., developed by the architect/engineer.

Bio-Based Materials: Products other than food that are biological products, renewable agricultural materials, or forestry materials derived from biomass. Plants and animals can be examples of bio-based materials; however, hide products such as leather and other animal skin materials are excluded in LEED calculations.

Biofuel: Fuels produced from organic material. Biofuel includes untreated wood waste, landfill gas, agricultural crops or waste, animal waste, and other types of organic waste.

Bioswale: A stormwater control feature that uses a combination of an engineered basin, soils, and vegetation.

Blackwater: Used water that has come into contact with waste. Thus, the water collected from the urinals and toilets can be classified as blackwater.

Blowdown: Removal of the cooling tower's water in order to minimize deposit of scales.

Brownfield Site: Previously developed sites that were contaminated with waste or pollution. A site that has an abandoned building with unknown contamination can also be classified as a brownfield site.

BUG Rating Method: A luminaire classification system that classifies a luminaire according to backlight, uplight, and glare.

Building Automation System (BAS): A computer-based monitoring system which can monitor, coordinate and control every individual building system.

Building Exterior: Defined as everything from the waterproofing membrane, inclusive of the waterproofing membrane.

Building Footprint: Describes the area that the building sits on.

Building Interior: Defined as everything inside the waterproofing membrane.

Built Environment: All the man-made surroundings that are needed for human activity, from roads, to buildings, to neighborhoods.

Charrettes: Intense workshops that are generally held at the beginning of the project and during project milestones. These workshops bring together the entire project team, facility manager, and all the stakeholders and outside experts/consultants. Charrettes allow brainstorming and collaboration among different disciplines and experiences.

Clean Waste: Materials that are left over from construction and demolition that are nonhazardous.

Climate Risk: The likelihood and potential severity of harm resulting from a climate-related hazard, taking into account the building's vulnerability, its occupants, and the surrounding context.

Climate Risk Assessment: A systematic evaluation of how identified climate hazards could affect a building's performance, safety, occupants, and the owner's investment over the project's expected service life.

Closed System: A system that does not produce any waste product by circulating the same medium continuously.

Commingled Waste: Single-streamed waste for recycling.

Commissioning (Cx): A systematic process carried out by skilled personnel to verify that a building's systems are designed, installed, and operating in accordance with the owner's project requirements (OPR), project goals, and design specifications. Building commissioning ensures that the project is designed and built as truly intended.

Compact Development: Development strategies that promote location-efficient neighborhoods with high development density, reducing greenhouse gas emissions and urban sprawl.

Compressed Workweek: A work arrangement in which employees work longer hours on fewer days in exchange for an additional day off during the workweek, reducing commuting frequency and associated fossil fuel usage.

Consensus Committees: Established by the USGBC Board of Directors to support the creation and updating of specific LEED rating systems. Each consensus committee is structured to represent different technical and market interests, and no single interest group may make up more than 25% of the voting membership.

Conventional Irrigation: Common system used for irrigation, such as irrigation through sprinkler heads above the ground.

Cradle-to-Cradle: The practice of designing materials to have "infinite" life cycles by recovering and recycling them into new products, creating a closed-loop system instead of waste.

Cradle-to-Grave: A life-cycle approach that evaluates a product from raw material extraction through manufacturing and use, all the way to final disposal.

Cradle-to-Gate Assessment: A partial life-cycle assessment that evaluates impacts from raw material extraction/harvesting through manufacturing, ending when the product leaves the factory gate (before transportation to the site, use, and end-of-life).

Decarbonization: The process of reducing or eliminating carbon emissions from building operations, embodied materials, refrigerants, and transportation. Decarbonization is the largest impact area in LEED v5, accounting for approximately 50% of available points.

Demand-Controlled Ventilation: A ventilation strategy that automatically adjusts outdoor airflow in real time using sensor signals. Most demand-controlled ventilation systems use indoor carbon dioxide levels as an indicator of occupancy—when carbon dioxide levels rise, the system increases ventilation, and when carbon dioxide levels drop, it reduces ventilation.

Demand Response (DR): An intentional reduction in the electricity usage in response to a demand response (DR) event or changes in the price of electricity.

Demand Response Event (Curtailment Event): The period that the utility company asks for a reduction in electricity usage from its program participants.

Dematerialization: The practice of intentionally reducing the total amount of materials used in a building without compromising its function or performance.

Development Density: A measure of the total building floor area or total number of dwelling units on a unit of land relative to the total buildable area on that land.

Development Footprint: Named for the sum of all the areas that are affected by the project's activity in the project site.

District Energy System (DES): A central energy conversion plant that provides thermal energy, shared by a group of buildings.

Diverse Uses: Publicly available businesses that provide daily need goods or services. According to USGBC, diverse uses do not include ATMs, vending machines, and touch screens.

Diversion Rate: Percentage of waste materials diverted from landfill.

Drip Irrigation Systems: Microirrigation systems that drip water to the roots of plants to minimize the use of irrigation water and fertilizers. They are the most water-efficient systems and have very short payback periods.

Dry Ponds (Detention Ponds): Excavated basins that temporarily hold and slow down stormwater during rain events, but remain dry most of the time. By detaining runoff, they reduce downstream flooding and allow water to infiltrate gradually into the ground as the pond drains.

EDUCATION @USGBC: Education portal of USGBC.

Electric Induction Cooking: A cooking method that uses a magnetic field to directly heat pots and pans, eliminating emissions from on-site combustion.

Electrification: The process of replacing non-electric energy sources with electricity.

Embodied Carbon: Measures the total greenhouse gas emissions associated with a product's entire life cycle—from raw material extraction and manufacturing to transportation, installation, and disposal.

Embodied Energy: The total energy consumed resulting from a product's manufacturing, transportation, installation, use, and disposal.

Emergent Properties: Emergence of certain properties in the systems as a result of interaction of individual elements.

Emissivity (Infrared or Thermal Emittance): A measure that shows how much heat or infrared radiation a material can shed back into the atmosphere.

Energy-Recovery Ventilation: A system in which exhaust air is used to condition incoming fresh air. During cold weather, heat from outgoing air is transferred to incoming air, while in hot weather, the process works in reverse to reduce the cooling load.

Energy Use Intensity (EUI): A measurement unit that describes the building's energy usage relative to its size.

Environmental Justice: The principle that every person deserves equal protection from environmental and health hazards, equal access to decision-making, and equitable access to a healthy and sustainable environment.

Environmental Product Declaration (EPD): Disclosure that reports a product's environmental impacts based on life-cycle assessment data. An EPD describes the environmental effects of raw material extraction, energy use, chemical composition, waste generation, and emissions to water, air, and soil throughout the product's sourcing and manufacturing process.

Environmental Tobacco Smoke (ETS): Smoke from another person's cigarette, cigar, pipe, or electronic smoking device. Breathing in ETS is known as passive smoking, secondhand smoking, or involuntary smoking.

Erosion and Sedimentation Control (ESC) Plan: A plan developed to prevent erosion, sedimentation, and stormwater pollution to the water bodies, wetlands, and the whole neighborhood.

Evapotranspiration: The return of water to the atmosphere through evaporation from soil surfaces and transpiration from plants.

Exemplary Performance Points: Additional points generally achieved by doubling the credit requirements or reaching the next incremental percentage threshold.

Extended Producer Responsibility (EPR): A product stewardship policy approach that holds companies producing consumer goods responsible for managing their own products and packaging when consumers are finished with them.

Extensive Vegetated Roofs: Types of roofs that do not include a variety of plants and require little maintenance. Their soil layer is thinner compared with the intensive roofs since they are more designed for the smaller-sized vegetation.

Facility Manager: The professional responsible for overseeing the day-to-day operations, maintenance, and management of a building to ensure it functions efficiently and meets the needs of its occupants.

Fenceline Community: A community located immediately adjacent to or in close proximity to a source of pollution, such as an industrial facility, highway, or waste site, and therefore disproportionately exposed to environmental and health hazards.

Floor-to-Area Ratio (FAR): A density measure calculated by dividing a building's total gross floor area by the total lot (site) area. For example, a 20,000 sq ft building on a 10,000 sq ft lot has an FAR of 2.0.

Flush-Out: The process of supplying large volumes of fresh air to the building before or during occupancy to remove contaminated air and establish the desired level of indoor air quality.

Formaldehyde: A hazardous volatile organic compound commonly found in adhesives used to manufacture composite wood products such as plywood and particleboard. It is one of the most critical indoor pollutants to control.

Functional Entry: Any building opening that is open and used by pedestrians during business hours.

Gallons per Flush (gpf): A water-use metric for flush fixtures (toilets and urinals) that shows how many gallons are used per flush.

Gallons per Minute (gpm): A water-use metric for flow fixtures (faucets and showerheads) that shows how many gallons flow per minute.

Glare: Excessive brightness, often from high-angle light, that can cause visual discomfort and reduce visibility.

Global Forest Watch: An online platform that international projects may use to understand tree coverage in the project neighborhood.

Graywater: Untreated household water that did not come into contact with toilet waste. Used water from bathtubs, showers, bathroom washbasins, and water from clothes washers and laundry tubs can be examples of graywater and may be used as a flush water in toilets or urinals. This definition can change depending on the local codes.

Green Building: The practice of creating structures and using processes that are environmentally responsible and resource-efficient throughout a building's life cycle, from siting to design, construction, operation, maintenance, renovation, and deconstruction. This practice expands and complements the classical building design concerns of economy, utility, durability, and comfort.

Green Classroom Professional (GCP) Certificate: Designed for educators and school facility professionals who want to gain a foundational understanding of green building principles in the context of a school setting, helping them identify what supports or impedes healthy, resource-efficient, and environmentally sustainable learning spaces.

Green Cleaning: The use of environmentally friendly products and cleaning practices to maintain buildings while minimizing harm to human health and the environment.

Green-e: The leading certification program for renewable energy generation in the United States.

Green Infrastructure: Infrastructure to direct the rainwater collected from the impervious surfaces to the vegetation and soil surfaces without routing them to the storm sewer system.

Greenfield: Land that has not been previously developed.

Greenwashing: Refers to the presentation of a product or a material as being more environmentally friendly than it actually is.

Grid-Interactive Building: A building that can adjust how and when it uses electricity depending on grid conditions, helping reduce pressure during periods of high demand, support renewable energy use, and contribute to decarbonization and resilience.

Halons: Chemicals used in fire suppression systems.

Hard Cost: Costs that directly pay for the physical construction of the building, including labor, construction materials, and equipment.

Hazard Assessment: An evaluation of which threats are most relevant to a project by analyzing climate projections alongside the specific conditions of the site.

Health Product Declaration (HPD): Disclosure that provides a product's material ingredients, list of potential chemicals, related concerns, and additional health information.

Heat Island Effect: The phenomenon in which dark-colored, nonreflective surfaces absorb heat during hot weather and release it into the atmosphere, raising local temperatures.

Heat Pumps: An energy-efficient system for heating and cooling spaces and for heating water, draws thermal energy out of water, air, or the ground into a refrigerant and then transfers that heat to another place.

Impervious Surface: A surface that contains less than 50% perviousness.

Inclusive Design (Universal Design): A design approach that aims to make buildings accessible to people of all ages, abilities, and demographics.

Indoor Air Quality Management Plan: A plan developed to protect the indoor air quality for construction workers and building occupants.

Infill Sites (Infill Developments): Sites in which at least 75% of the total area was either previously developed or already being used for other purposes in urban areas.

Inherently Nonemitting Materials: Materials with very low or no VOC content.

Integrated Pest Management (IPM): A sustainable approach that combines knowledge about pests, nature, pest prevention, and control methods that minimize pest infestation and damage while minimizing hazards to the building occupants, the property itself, and the environment.

Integrated Process: An approach that emphasizes early and ongoing coordination among the project team. Owners, architects, and engineers (structural, electrical, mechanical), along with facility managers and other key parties, work collaboratively starting in the predesign phase to align goals and make better decisions.

Intensive Vegetated Roofs: Green roofs with deeper growing media that can support a wider variety of plants (including larger plants), but typically require more structural support and maintenance than extensive roofs.

Intergovernmental Panel on Climate Change (IPCC): The United Nations body responsible for assessing the science related to climate change.

Invasive Plants: Plants that spread and damage the environment by taking over the adjacent existing native and adapted plants.

Islanding: The capability of an on-site energy system to disconnect from the main grid and operate independently during an outage, allowing the building to safely use its own energy resources and continue providing essential services.

Land Trust: A nonprofit organization that works on conserving lands.

Landscape Water Requirement (LWR): The amount of water that the landscape of the site will require during the site's peak watering month.

LEED AP with Specialty: LEED credential, created for professionals with advanced knowledge in green building practices and specialized in a particular LEED rating system.

LEED Campus Program: A program designed for multiple buildings on a shared site under the control of a single entity. It allows project teams to document shared site-related credits once and apply them across all individual building certifications on the campus.

LEED Combined Review: A type of LEED certification review in which the documentation for all design and construction prerequisites/credits is submitted for review at the end of the construction phase (for LEED BD+C and LEED ID+C rating systems).

LEED Fellow: LEED credential created to designate the most exceptional professionals in the green building industry.

LEED Green Associate: LEED credential created for professionals with a proven, up-to-date understanding of green building principles and practices.

LEED Green Rater: Professional who provides in-field verification to LEED Residential projects.

LEED Impact Categories: The high-level outcomes that LEED projects are designed to support. Every LEED prerequisite and credit aligns with one or more impact categories: Decarbonization, Quality of Life, and Ecological Conservation and Restoration.

LEED Interpretation: A precedent-setting official answer to a formal inquiry about the technical details of a LEED prerequisite, credit, or MPR. Unlike other forms of guidance, LEED interpretations apply to all LEED projects, not just the one that requested it. LEED interpretations are administered by USGBC and published in the Addenda database.

LEED Project Administrator: The primary project contact for GBCI. The LEED Project Administrator is the team member who acts as a project manager, overseeing the LEED project and organizing team members for certain tasks, credits, or prerequisites.

LEED Project Boundary: Portion of the site submitted for LEED certification. Defined by the platted property line of the project, including all land and water within it.

LEED Recertification: Required for LEED O+M projects to maintain their certification every 3 years. Projects certified under other LEED rating systems do not require recertification.

LEED Scorecard: Allows project teams to select the credits they plan to pursue and determine whether the total expected points will be sufficient to achieve the targeted level of LEED certification.

LEED Split Review: A type of LEED certification review in which the design prerequisites/credits are submitted for review during the design phase, and both the additional design prerequisites/credits and all the construction prerequisites/credits are submitted at the end of the construction phase (for LEED BD+C and LEED ID+C rating systems).

LEED Steering Committee (LSC): A standing committee established by the USGBC Board of Directors. The LSC sets the strategic direction for LEED development, ensures consistency among all LEED rating systems, and reviews recommendations from other committees to confirm alignment with the overall LEED framework.

LEED Volume Program: A certification approach for organizations planning to certify a large number of new construction buildings or interior spaces based on a common prototype, streamlining certification by focusing on similarities in building design and construction practices regardless of location.

Leverage Point: The point where any action taken in the system can bring about significant results.

Life-Cycle Approach: Evaluates the entire life cycle of a project, product, or service.

Life-Cycle Assessment (LCA): A method that quantitatively evaluates a product's environmental impacts across its life cycle. A cradle-to-grave, or cradle-to-cradle, approach is used in LCA, and the total energy use and other environmental consequences resulting from the creation of that material are calculated.

Life-cycle Costing (LCC): A method that estimates a product's total cost over its life, including initial cost, operation, maintenance, service/repairs, and end-of-life disposal.

Light Shelves: Horizontal, light-reflecting overhangs that are positioned to reflect the daylight into the desired area of the building.

Light Tubes: Reflective tubular devices that capture daylight at the roof and channel it into interior spaces to provide natural light deeper inside a building.

Lighting Power Density (Connected Lighting Power): The amount of power used per square foot or square meter when all lighting is turned on. Appropriate levels vary by space type.

Linear Approach: A traditional project delivery approach where each team member works separately in sequence, then hands their work off to the next person, with limited collaboration across disciplines.

Load Shedding: The intentional action by a power utility to reduce the load on the power system in order to prevent a total failure of the system.

Load Shifting: Storing energy during off-peak hours and using it later during peak-demand hours to reduce peak loads and electricity costs.

Location Efficiency: A measure of how well a project's surroundings reduce the need for driving by providing easy access to transit, services, and amenities within walking or cycling distance.

Low-Impact Development (LID): A rainwater management strategy that mimics natural site hydrology by managing stormwater close to where it falls. LID strategies include decreasing impervious surfaces and increasing vegetation on-site. LID is used to slow down the flow of rainwater out of site and ensures that the rainwater does not get contaminated.

Makeup Water: Water used to replace the lost water in open systems.

Masking Systems: Equipment used to reduce the background noise in spaces.

Material Circularity: A system in which resources are continuously reused and recycled, reducing dependence on virgin materials and minimizing waste.

Material Efficiency: Using less material to perform the same function or achieve the same results.

Minimum Efficiency Reporting Value (MERV): A rating system that measures how effectively an air filter removes particles from the air. Air filters with higher MERV values will remove more particles from the air, but more energy will be consumed in passing the air through them.

Minimum Program Requirements (MPRs): The minimum standards/characteristics that every LEED project should meet to pursue LEED certification. MPRs state whether a project can pursue LEED certification or not. They provide guidance on the types of projects that are eligible for LEED certification, protect the integrity of the LEED program, and reduce the number of issues that arise during the certification process.

Monitoring-Based Commissioning (MBCx): The process of using software to monitor real-time data from building automation systems and meters, enabling building operators to detect faulty equipment, identify unusual energy consumption patterns, and address issues as they occur.

Monitoring-Based Commissioning Provider (MBCxP): A contracted third party or qualified in-house staff person responsible for leading the monitoring-based commissioning process.

Mulching: A protective layer applied to the surface of soil that will help to keep the roots of the plants cool and therefore prevent evaporation.

Native Plants (Indigenous Plants): Plants that occur and develop naturally in a specific location.

Natural Refrigerants: Refrigerants that occur in nature's biological and chemical cycles without human involvement, such as carbon dioxide (CO_2), water (H_2O), ammonia (NH_3), air, and hydrocarbons such as propane, ethane, and butane.

Negative Feedback Loops: A feedback loop where a change causes a response in the opposite direction to bring the system back toward balance. If a room gets warmer than the set temperature, the thermostat will send a signal to the air conditioning, and the air conditioning will stop blowing warm air.

Net-Zero Energy Project: A project that produces at least as much renewable energy as it consumes on an annual basis.

Nonpoint Source Pollution: Type of pollution in which its source cannot be identified and which generally results from multiple sources.

Nonpotable Water: Water that does not meet the human consumption standards.

Nonprocess Energy (Regulated Energy): The energy consumed by the items that are used to condition spaces and maintain comfort and amenities for building occupants.

On-Site Water Retention: The practice of collecting and retaining rainwater that falls onto pervious surfaces on the project site.

Open Systems: Systems that constantly consume other items, use them, and produce waste at the end.

Open-Grid Pavement: A pavement system with at least 50% unbound.

Operational Carbon: Emissions produced by running the building on a daily basis, such as energy used for heating, cooling, lighting, and powering equipment.

Overage: Extra material ordered as a precaution during construction that often goes unused. Minimizing overage through careful design and procurement reduces waste.

Owner's Project Requirements (OPR): A document that outlines the owner's vision and performance requirements for a project, including the LEED objectives the project needs to accomplish.

Particulate Matter (PM): A mix of tiny solid particles and liquid droplets floating in the air, most of which can only be detected using a microscope. PM can carry contaminants such as heavy metals, mold spores, dust mites, combustion by-products, pest droppings, and chemicals from pesticides, cleaning products, or building materials.

Passive Daylighting: A system that collects sunlight using static, nonmoving elements such as windows, glass doors, skylights, light tubes, and light shelves.

Passive Survivability: A building's ability to keep occupants safe by maintaining critical life-support conditions even when services such as power, heating fuel, or water are lost.

Pilot Credits: Credits being tested across all rating system types and credit categories and include credits proposed for the next version of LEED.

Pilot Alternative Compliance Paths (Pilot ACPs): Compliance paths that USGBC is considering for inclusion in future versions of LEED, which can optionally be used in place of existing LEED credits.

Plug Loads (Receptacle Load): Represents the electrical use by all the equipment that is connected to the electrical system via electrical receptacles.

Positive Feedback Loop: A feedback loop where a change causes more change in the same direction—A produces B, and B produces more of A. For example, in an interest-earning savings account, the balance earns interest, and as the balance grows, it earns even more interest, which accelerates the growth further.

Postconsumer Recycled Content: The recycled content of a used material. For example, recyclable printer paper can be sent to recycling after being used and can become a part of new printer paper. Other types of materials with postconsumer recycled content can be aluminum cans, water bottles, most glass, wood and steel products, newspapers, and more.

Potable Water: Water that is approved for human use that meets or exceeds US Environmental Protection Agency drinking water quality standards (or a local equivalent outside the United States.)

Preconsumer Recycled Content: The content of a material that is recycled before reaching any consumer. An example is sawdust generated during the manufacturing of a wood product that is recycled to be used inside an MDF board (medium-density fiberboard).

Preferred Parking: Parking spaces that are closest to the main entrance of a building.

Prerequisites: The minimum requirements that all buildings under a certain rating system must meet in order to achieve LEED certification.

Previously Developed Site: A site that contains at least 75% previously developed land.

Prime Farmland: Land that is used for producing food, feed, forage, fiber, and oilseed crops or is available for these uses, as determined by the U.S. Department of Agriculture's Natural Resources Conservation Service.

Prius Effect: The idea that people change their behavior when they can see real-time feedback. In other words, people can respond to something only if they have immediate information about it.

Process Water: Water used by mechanical or other types of systems in buildings such as cooling towers or medical equipment in hospitals.

Project Priorities Library: A USGBC resource where new credits can be added as they are developed.

Project-Type Credits: Credits designated by USGBC to address the specific, unique needs of a particular adaptation or building application, such as the Places of Respite credit available only to healthcare projects.

Radon: A radioactive gas naturally found in soils, rocks, and water bodies. Exposure to radon is harmful to human health and is the second leading cause of lung cancer after smoking.

Rainwater Harvesting: An aspect of rainwater management that collects and filters the rainwater to be reused as an alternative to potable water.

Rapidly Renewable Materials: Natural materials that can replenish within 10 years.

Reclaimed Water: Former blackwater that has been treated and purified for reuse.

Reference Soil: The site's original, undisturbed native soil—the soil condition that existed before development and that is used as the baseline for comparison (for example, when evaluating soil restoration or soil health).

Refrigerant: Substances used to transfer heat.

Regenerative Design: A design approach where the project is intended to give back more than it takes—producing more energy than it consumes, generating little to no waste, and actively restoring ecosystems and habitats that development has degraded. Excess energy produced on site can be returned to the utility grid, making the building a contributor to the energy supply rather than just a consumer.

Regional Alternative Compliance Paths (Regional ACPs): Credit requirements that may optionally be applied in specific regions of the world to address local standards and conditions.

Renewable Energy Certificates (RECs): Tradable, intangible instruments representing the environmental and other non-power attributes of renewable electricity generation. One REC generally represents 1 megawatt-hour (MWh) of electricity generated from an eligible renewable resource and delivered to the grid.

Renewable Energy: A type of energy that is derived from renewable sources. Renewable energy includes solar, wind, wave, biomass, and geothermal power, plus certain forms of hydropower.

Resilience: A building's ability to prepare for, withstand, respond to, and recover from disruptive events.

Retrocommissioning: A systematic process for analyzing and optimizing building system performance involving a commissioning provider. It includes testing of building systems and documentation of any measures needed to improve performance.

Reverberation Time: The time span between when a sound is produced and when it dies away.

Scope 1 Energy: Direct energy generated from sources owned or controlled by the organization, such as on-site fuel combustion.

Scope 2 Energy: Indirect energy purchased from external sources, such as electricity, steam, or district heating and cooling.

Scope 3 Energy: Indirect energy from sources not owned or directly controlled by the organization, such as employee commuting, business travel, or supply chain activities.

Sensitive Lands: Ecologically sensitive areas such as prime farmland, floodplain, habitat, water bodies, or wetland.

Site Assessment: A part of the integrative process, which clearly shows the project teams the properties of the site, including its topography, hydrology, climate, soil types, water availability, and human health effects.

Smart Growth: A neighborhood development approach that protects undeveloped lands and contributes to project development in locations near jobs, schools, shops, and other diverse uses.

Social Determinants of Health: The conditions in which people are born, live, and work that play a significant role in shaping well-being, including factors such as access to clean air, safe housing, green spaces, nutritious food, and quality education.

Soft Cost: Costs needed to plan, design, and manage a project that do not physically become part of the building. Soft costs include items such as design and consulting fees, project management and supervision, permits, legal fees, financing, insurance, and taxes.

Solar Reflectance (SR): Shows the solar energy that is reflected by a surface on a scale of 0 to 1. A black surface will have a SR of 0 while a white surface will have a SR of 1.

Solar Reflectance Index (SRI): Indicates a material's ability to stay cool by reflecting solar radiation and emitting thermal radiation. Thus, both the solar reflectance and emissivity of a material will be combined to rank the material.

Source Control: The practice of identifying sources of indoor contaminants and taking measures to prevent them from reaching the indoor environment.

Source Reduction: The exact sizing of materials through prefabrication, modular construction, or similar methods to prevent waste generation on-site. It also covers the use of products with less packaging.

Suburban Sprawl: The expansion of populations away from central urban areas into low-density areas.

Sufficiency: A set of measures and practices that avoid unnecessary demand for energy, materials, land, and water while still delivering human well-being within planetary boundaries.

Sustainable Development: Meeting the needs of the present without compromising the ability of future generations to meet their own needs.

Systems Thinking: Refers to the understanding of each and every system of a building while also understanding their relationships and looking at the project as a whole.

Technical Advisory Groups (TAGs): Committees that advise the LEED Steering Committee and consensus committees on specific technical aspects of LEED, such as materials or indoor air quality.

Thermal Bridging: Areas where conductive materials within the building envelope permit unwanted heat transfer in both cold and hot weather conditions.

Thermal Energy Storage: A system that stores heating or cooling capacity for later use, typically in an insulated tank, rather than storing electricity.

Thermal Mass: The ability of a material to absorb and store energy. High-density materials such as concrete and brick require significant amounts of heat to change temperature.

Tier 1 Renewable Energy: On-site renewable energy generation, including systems located on the project site, on the surrounding campus, or on the site of a social impact project where the system is donated and ownership is transferred to the recipient. Eligible resources include solar electric (photovoltaic), solar thermal, wind, and recovered heat from municipal wastewater.

Tier 2 Renewable Energy: New off-site renewable electricity from generation assets contracted to be operational within two years of occupancy or within five years of their commercial operations date.

Tier 3 Renewable Energy: Off-site renewable electricity or fuels that are Green-e Energy certified or equivalent.

Total Volatile Organic Compounds (TVOC): The combined concentration of many VOCs reported as a single value.

Transportation Demand Management (TDM): A set of rules and strategies designed to reduce single-occupancy vehicle trips, including measures such as public transit incentives, carpool programs, employer commuter benefits, limits on parking supply, and policies that encourage walking, biking, and public transit.

Tree Equity Score: A tool that highlights inequities in tree cover and identifies where trees are needed most in each community, helping address disparities by focusing resources in neighborhoods with the highest need.

Triple Bottom Line (Three Pillars of Sustainability): A framework that evaluates a project's success across three dimensions: economic prosperity (profit), social equity and justice (people), and environmental stewardship (planet).

Uplight: Light emitted above the horizontal plane that contributes to sky glow.

Volatile Organic Compounds (VOCs): Chemicals that evaporate at room temperature and can be released from building products such as paints, adhesives, sealants, flooring, and furniture. Some VOCs can cause short-term effects like headaches or long-term health problems including cancer.

Walk Score: A measure of a location's walkability on a scale of 0 to 100, based on walking routes to nearby amenities. A higher score indicates more amenities and services within a short walking distance.

Wet Pond (Retention Pond): Pond designed to hold a specific amount of water indefinitely.

Working Groups: Groups established by the LEED Steering Committee or a consensus committee to complete a specific task with a clearly defined scope and timeline.

Xeriscaping: Type of landscaping that does not need any irrigation.

Zero-Lot Project: A type of project in which the building footprint extends all the way to one or more property (lot) lines, leaving little or no space for setbacks on those sides, except for limited access areas (e.g., entryways, walkways, drive aisles, or service access).